Alexander King

Alexander King

BUSES ANNUAL
1967

A Saurer coach of the Swiss Post Office negotiating the Klausen Pass between Altdorf and Linthal in the 1930s.

BUSES ANNUAL 1967

Edited by
R. A. Smith

LONDON
IAN ALLAN

© Ian Allan 1967

Published by Ian Allan Ltd., Shepperton, Surrey, and printed in the
United Kingdom by Netherwood Dalton & Co. Ltd.

CONTENTS

		Page
THE GREATEST TOUR OF ALL	F. W. York	7
HAY-DAY FOR A GUY	G. R. Mills	18
CO-ORDINATION	Charles F. Klapper	25
THE RAMBLINGS OF AN ENTHUSIAST	Oscar Alison	32
LANCASHIRE-SCOTTISH EXPRESS SERVICES	A. J. Douglas	46
"NULLI SECUNDUS"	G. J. Robbins	58
THE POSTAL BUSES OF SWITZERLAND	J. Graeme Bruce	71
THE LEYLAND STORY: 1926-1942	A. Alan Townsin	78
THE SECOND DECK	Charles F. Klapper	94

On the cover. Cardiff has one of the best-sited bus stations in the country, on the doorstep of Cardiff General Western Region station. The bus in the foreground is a Western Welsh Weymann—bodied 65-seat PD2A/27 Leyland Titan, delivered in 1963.

[C. F. Klapper

The three London buses which toured North America in 1952 on the eastward part of their journey following the American River through the Sierra Nevada mountains.

The Greatest Tour of all

by F. W. York

IN these days when the nation as a whole is drawn ever increasingly towards the realisation of the importance of healthy export markets, the news that London Transport is making available a double-deck bus to assist in the staging of a trade fair or exhibition to show the British way of life in some distant part of the world is only sufficient to warrant a small paragraph in the centre pages of the daily newspapers. Like 'dog bites man' it has happened so often over the past few years that it borders on the commonplace. The pioneers of these overseas visits deserve to take their rightful place within the pages of transport history.

It is difficult to say exactly when London buses went abroad in support of national interests. Some might be tempted to quote the fleets that found their way to the fields of Flanders during the first World War, but as this was for a very different purpose—nothing to do with trade—it has no bearing on this account. It is a fact that in the early years after that war the London General Omnibus Company allowed an S-type double-decker to be placed on exhibition at Calcutta; the full details of the visit seem to have passed away in the flow of time, but it seems that it took place in the winter of 1923, a point upon which readers might have more complete knowledge. Furthermore, it is not known if it was a static exhibit, or whether it demonstrated its capabilities. It is interesting to speculate that this bus might have been the inspiration for the rather ungainly double-deckers that worked on the streets of Calcutta during the early 1930's, with bodies that were so obviously of local construction being carried on a selection of unlikely chassis—one hundred passengers on a Commer, for instance!

Without question, the fact that overseas ventures undertaken by London buses are regarded as being a success almost before they leave these shores is almost wholly due to three of their number and the magnificent performance they gave during a twelve thousand mile tour through the United States

Cover of the British Travel Association brochure and map of the tour drawn by Emett, the well known artist.

[E. T. Bonney collection

and Canada during the spring and summer of 1952. The decade following the conclusion of the second World War was one in which the country had to make a painfully slow recovery from the austerity imposed by wartime conditions. The government of the day decided to show the world in general just what progress was being made, and, against a barrage of cynical comment, made plans to hold the Festival of Britain in 1951. The success that it enjoyed dumbfounded the critics, and at the same time demonstrated what an important factor in the national economy tourism could be. In addition to the work undertaken at home, such as the construction of new tramways on and around the South Bank site of the Festival and the provision of special bus services for visitors, London Transport had also made available four RTs, whose task it had been to carry out a publicity tour of 5,000 miles, taking in eight European countries. It was a complete success, and the fact that it was wholly trouble free inspired the British Travel and Holidays Association, as it then was, to prepare plans for a 'Come to Britain' campaign aimed initially at the United States of America. The Association had no doubt felt that a static exhibition or fair held at one of the principal centres of population would not have the impact that was so essential if the message was to be projected throughout the country. Therefore, something that was mobile and "different", and yet essentially British, had to be sought. To the American tourist the British Isles means first and foremost the history that is London, and what could be more characteristic of that city than the big red buses that predominated the scene? If a campaign using buses in Europe could be a success, why not attempt the same thing in America?

Those of us who see and use double-deck buses every day of our lives would find it difficult to appreciate the reactions of some other person facing one for the very first time. To speak from personal experience, I lived for some years in a part of the world where the single-decker was the only type in use. Then, when a double-decker arrived on loan for evaluation purposes it seemed to be quite enormous, and in a way, a rather wonderful piece of engineering skill. One of my favourite comic post cards shows two drunken passengers seated in the upper saloon of a bus, with one saying to the other "It's not safe up here, there's no driver": such disquiet was clearly to be seen on the faces of those local inhabitants hardy enough to sample an upper-deck ride on this particular demonstrator.

Once the media of the campaign had been decided, a great deal of advance planning had to be done by the Association, with the full support of the British Embassy in Washington, working in close co-operation with several American government departments and the police forces through whose territory the proposed route lay. Clearances would be of prime importance and had to be most

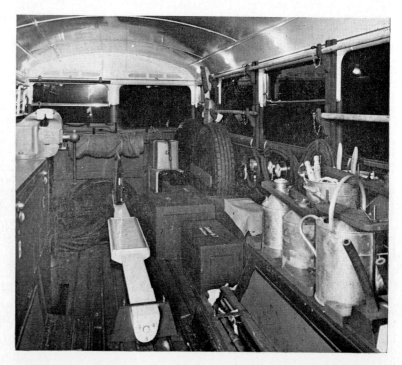

The lower deck of RT2775, with blacked-out windows, was fitted out as a mobile workshop and engineer's store; spare wheels, hoses, brushes, brooms, buckets, chocks and blocks, a bench with vice, and a complete bus stop (in floor centre) were carried.
[London Transport]

carefully considered, for when railway bridges were built nobody thought that double-decker buses would have to pass beneath them. In most States the minimum statutory clearance beneath bridges was about 14 feet, but in some States it was less. Furthermore, whilst the main highway network is second to none, the secondary roads, over which diversions might have to be made, could be of a most indifferent nature—little more than dirt roads, in fact.

As planned, the tour was to cover 10,000 miles, and to find the finance was thought to be a problem. As it happened, industrial concerns were quick to come forward and take the advertising spaces that are normally offered on London's buses. The London Transport Executive happily made available the vehicles and the crews to man them. It had been decided that three buses would be needed, and those selected were all brand new; two were RTs, of which one, RT 2776, was to be used to give rides at the various places where halts were to be made. All vehicles were standard machines, but in the case of RT 2776 it was felt that, bearing in mind the high temperatures through which it was to be called upon to operate, always with a full passenger load, more extensive ventilation should be provided. The vents in the front dome and the projecting cowl beneath the canopy were to become features of easy recognition. American law demands that marker lights be fitted, thus adding another external feature. Headlamps and fog lamps for the trio had to be specially manufactured to meet the requirements of American law and the manufacturers of this equipment, Messrs. CAV also fitted their pilot injection system to RT 2776. The other was RT 2775, whose rôle was to be that of combined crew quarters, workshop and literature store—although so great was the amount of publicity material to be handed out, both on the subject of the British Isles in general and the buses themselves in particular that stockpiles of material had to be sent in advance to selected points en route, to replenish 2775's supply. The pamphlet produced by London Transport described the workings of an RT and then went on to explain why it was that the British as a whole preferred double-deckers.

The trio was completed by RTL 1307, which was fitted out as a mobile exhibition, the displays showing various facets of British life. Dispensing information on the vehicle were young ladies from the Association's office in New York, whose scarves and certain items of luggage proudly bore the LTE's 'Bulls-Eye' motif. The body was by Weymann's, and was of interest in that this was the first RTL chassis to be thus equipped. In attendance with the buses was an Austin 'Atlantic' car for use by the convoy leader in his capacity as pathfinder, a Leyland 'Comet' lorry to carry heavy spares—and at the same time exhibit a diesel engine—and a Fordson 'Thames' 10-cwt. van fitted with a generator with which to provide current for lighting the displays in RTL 1307.

The impossible situations that existed upon the railways created by Rowland Emett in his famous cartoons were much appreciated by the Americans, and his work was largely featured in much of the advance publicity work. For example, here came the three buses in V-formation, the leader having Big

Interior of the upper (top) and lower decks of RTL1307, the exhibition bus.

[London Transport

Ben growing from the top deck, whilst the others had the Eros Statue and Nelson's Column respectively. Monacled gentleman waving Union Jacks gave icy smiles from the upper windows, whilst conductors manfully threw out clouds of tickets which, on another poster, fell neatly into the outstretched hand of the Statue of Liberty.

Whilst the convoy was being assembled the team that was to act in an almost ambassadorial capacity was undergoing a course of intensive training, not the least important aspect of which being the art of driving on the right-hand side of the road. All were hand picked men—bus drivers or engineering staff—the best available, and years of driving on London's streets must have made the 'keep left' rule a sixth sense that was not easy to break. There were also the American traffic laws and Highway Code with which to become familiar.

Events moved smoothly and on February 29, 1952, the convoy was drawn up on the Horse Guards Parade for an official inspection and send-off. Heading north, the next major point was the assembly at the dockside in Liverpool, and all were alongside the Cunader 'Parthia' by March 5. In order to maintain the high standard of appearance, they were shipped below decks, which called for a high measure of precision from the operator of the hugh floating crane; loading was completed without mishap, despite the fact that clearance at the hatchways was only six inches. Britain was left behind on March 8 and eight days later 'Parthia' docked at New York.

The programme called for the longest stay of the tour to be made at New York, between March 18 and 25, but even at that early stage it was possible for a slight snag to arise. The day following docking was St. Patrick's Day, and in view of the high percentage of Irishmen in the dock labour force it was thought wise to leave the buses where they were until after the festivities—not for any ulterior motive, but because there might be insufficient men at work to move them. The resulting spare time was not wasted, for the crews had to face the American driving test. By the noon of the 18th the buses had been offloaded and were ready for their first run on American soil—the drivers must have been apprehensive in the extreme. With an escort of police outriders announcing their advent with wailing sirens, the convoy left the docks and passed through the business heart of the city to reach the City Hall and the first of many official welcomes.

For the first three decades of the present century local passenger transport needs in the United States were met by extensive tramway networks. Towns were linked by interurban lines, a cross between a high-speed tramway and light railway that had no direct comparison in this country. Such was the strength of these traction empires that the true motor bus had a very late introduction, and then only in a limited way. The domestic manufacturer did not consider the large scale production of chassis designed specifically for bus work until the early 1920's and rather expected the prospective operator to adapt an established automobile chassis to his needs. When the Fifth Avenue Coach Company, of New York, wanted to introduce a motor bus as an experiment, with a view to changing over from horse traction, it was obliged to import from France a de Dion-Bouton double-decker. This proved to be an immediate success and in the following year, 1906, fourteen more similar chassis were purchased, the bodies being provided by the world famous tramcar constructors, J. G. Brill & Company of Philadelphia. Within the next twelve months no less than twenty more chassis were purchased, and the horse bus facilities were completely withdrawn. The Company was to gain the reputation of being the only American concern to develop the double-decker, having a fleet of vehicles of quite modern appearance in service until a year or two ago. They were ideal for use by tourists. Other cities tried double-deckers in varying degrees— a route along the shores of Lake Michigan that was operated by the Chicago Motor Coach Company again had such a great tourist traffic that a large fleet of double-deckers was used; Philadelphia, Clevelend, Baltimore, Washington, Kansas City, Atlanta, Los Angeles and Pittsburg also figured on the list of users. The last named is of particular interest in that at least two AECs were purchased; these were standard LGOC NS buses, even to the red livery and the 'General' fleetname—the owners' title, 'Pittsburg Motor Coach Company' was carried in large letters in between decks. Of the known two that were delivered, one was open top and the other enclosed, but both were fitted with solid tyres and the only concession to their American owner was that a right-hand entrance was provided, meaning that to English eyes the stairway was reversed. Except in New York and Philadelphia, the double-decker was extinct by 1940, and even those two bastions have now fallen.

To return to 1952—considerable thought had obviously been given to indicator displays On RT 2776 a modified display for route 11 was shown throughout the tour; the intermediate blind carried the names of all those places which Americans like to think of as being London—Buckingham Palace Rd. Westminster Abbey, Whitehall, Charing X, Strand, Fleet St., St. Pauls, Bank—and yet are located along the normal route of service 11. On the other two, a beflaged motif replaced the service number, the intermediate blind read 'Greetings From Britain' whilst the final destination showed the name of the town that was being visited prefixed by the word 'To', i.e. 'Greetings from Britain to Albuquerque'. The side box carried the simple message 'Come to Britain' whilst the canopy route number position was occupied by an illuminated Union Jack.

The initial success enjoyed at New York augered well for the rest of the journey but even so there were some dissenting voices to be heard, but in the friendliest way. Their argument was that, sound as the idea for the venture might be, the combination of climatic conditions and sheer distance involved must surely invite disaster. Not to be dissuaded by such pessimism. the convoy left for Washington, arriving for a three day visit on March 30, having spent two whole days at Philadelphia and single days at other towns on the way. (It is not proposed to list all the places at which stops were made, as this would tend to turn these notes into a gazetteer).

A feature of transport in the Federal capital was that the tramways were constructed with the conduit system of current collection, which must have made the London Transport men feel at home, but that was the only comparison that could be made between the two cities; whilst Washington could boast modern cars that looked as though they would give decades more service—which, as it turned out they didn't, at least, not in their own city— the breakers were very busy in the Charlton

RT2775, RTL1307 and RT2776 under the shadow of the Capitol in Washington, D.C.

[E. T. Bonny collection

tramatorium. By one of those strange coincidences that history sometimes produces Washington trams and London buses were to meet again, but in Sarajevo, Yugoslavia; the trams were purchased to operate a new line, and at the same time replace an established bus service which was using ex-London Transport Leyland PD1s of the post war STD class!

In all, twenty-eight states of the Union were passed through and 56 towns and cities visited, and a routine for arrival was soon developed. This was evolved largely at the request of the police, probably as a result of an incident during RT 2776's first solo trip in New York when the driver suddenly found the vehicle roof uncomfortably near the underside of part of the New York El (elevated railway) structure. Unable to go forward because of the bridge and unable to reverse because of following traffic, RT 2776 caused quite a traffic tangle before extricating herself. Therefore, while a police posse helped to create the VIP image, the police were also helping themselves by being 'on the spot' when difficulties arose; in fact, the tour leader was advised by the New York police not to cross any state border until the new escort from the next State arrived, and the wisdom of this advice was proved on several occasions when a low bridge or other overhead obstructions necessitated a detour—the longest added about 100 miles to one day's run—and the knowledge of the local police was vital in finding the best way round. The police, through radio communication, were also able to warn the next town of the delay.

Other sources of delay were junctions controlled by automatic signals, for often the traffic lights were suspended by cables over the junction; projecting advertisement hoardings could also be the source of nightmares as regards clearances, and one of the drivers remarked, at the end of the journey, that when back on normal duty it would be a long time before he stopped instinctively looking up to gauge the height of his bus in ratio, for example, to telephone wires carried across his road! At each State border it was usual to find a group of civic dignitaries waiting who, along with a police escort, conducted the convoy to the next town that was to be visited. At the City Hall a scroll containing a message of greeting from the Lord Mayor of London was handed to the local Mayor. At the same time, the exhibition bus was made ready while the crews were busy with cloths, eliminating dust traces of the journey. As soon as the speeches had concluded, there was a rush for rides on RT 2776, and one of the most difficult tasks of the whole tour was to instruct intending passengers in the noble art of queuing. The 'conductors' kept up a steady flow of 'Cockney chat' whilst issuing

Above. The tour leader checking the clearance under a lowish bridge. **Right.** A close shave as RT2775 just clears another low bridge on the way to Baltimore—note the rear marker lights on the cream band between decks.

[E. T. Bonny collection

to each passenger a special free commemorative ticket which again was a platform from which to extend both greetings and an invitation to visit Britain.

Moving ever westward, the crowded industrial eastern seaboard was replaced by fertile plains, and destination blinds began to show names that conjured up dreams of the Wild West—Kansas City, Oklahoma City, Santa Fe. Of all the geographical conditions to be faced, amongst the more severe was the run across the desert between Dallas and Santa Fe. At the start is to be found a notice reading 'Take plenty of water because the nearest is 1,000 miles away'. The truth of this message could be seen along the way in the form of cars that had been caught in sandstorms, blown over and wrecked. Although sandstorms were encountered, no damage to vehicles is recorded and the crates of spares that were carried were brought home unopened; indeed, it has been said the buses returned to Britain with the original London air in their tyres! Following the desert, other natural wonders were to be found, such as the multi-coloured sands of the Painted Desert and the marble-like texture of fallen trees in the Petrified Forest.

The Pacific Ocean was reached at Los Angeles, where it was only natural that RT 2776 should visit various film studios; a great number of publicity photographs was taken, one of the most widely reproduced showed Hollywood's version of a typical English policeman complete with thick horn rimmed spectacles! Only three days of the week spent in the City of the Angels were devoted to exhorting the local populace to 'come to Britain'; the crews spent three days having a

Above. Front and back of the special free tickets (actual size) issued to riders in Los Angeles.

Below. A broken axle on the trailer moving this house caused a two hour delay on the road to Roswell, New Mexico.

[E. T. Bonny collection

well earned rest—the only long break during the whole of the 17-week tour of the States, although they had odd days on five other

Maintenance day (May 9) at the General Petroleum Corporation depot in Los Angeles.
E. T. Bonny collection

occasions. The remaining day was required for the vital task of vehicle maintenance; only two other days were set aside in the original itinerary for mechanical check-up, but more were added, of course, when Canada was brought into the programme.

At San Francisco the cavalcade arrived at the western extremity of the tour, with two months having been spent most effectively in the crossing of a continent. With words it is difficult to convey distance, for not only are there the climatic changes to consider, but also the battle against time itself. When four o'clock chimes in New York (Eastern Standard Time) it is three o'clock in Kansas City (Central Standard Time), two o'clock in Albuquerque (Mountain Standard Time) and only one o'clock in San Francisco (Pacific Standard Time). A regular daily service to New York is provided by Continental Trailways, who term it 'The Golden Nuggett Route', the through coach being called 'The California Nuggett' when westbound, in the reverse direction the name being changed to 'The Manhatten Nuggett'. Passengers should be equipped not only with a timetable but also a calendar; for example, leaving San Francisco at 19:30 on Monday, then Salt Lake City is passed by 14:30 on the following day, Omaha at 15:45 on Wednesday, Pittsburg by 21:30 on Thursday and arrival in New York is 06:30 on Friday. Fortunately, a symbol in the timetable indicates that the through vehicles are 'Restroom Equipped Buses'. (By and large, the route that has just been described was the one to be followed by the London trio on their return across the continent).

May 15 to 17 was spent at San Francisco, where the London trio added to an already diverse local transport scene. As might be expected, they were often compared with the world famous cable cars, but the city can also boast one of the few remaining American tramway systems, which, although much reduced from its maximum mileage, embodies many strikingly modern features, such as considerable lengths of subway. Such tram routes that had been replaced provided the basis for an extensive trolley bus network—after the cable cars, this type of vehicle was found to be most effective for dealing with the steep hills of the city. Then, the municipality also ran a fleet of motor buses.

Turning eastwards again, the high spot of the journey was reached, in the most literal sense, for the Sierra Nevada Mountains forced the convoy to an altitude of 9,000 feet. The drivers were urged to the limits of their concentration, otherwise a moment of inattention could have sent a bus crashing down the sheer side of a mountain. At the particular time, 7-ft. snow drifts were being reported, but later the same day the descent brought a return to warmer temperatures. Nature again presented a barrier in the form of an arch of living rock over the road near Lake Tahoe, and clearance was obtained with only an inch to spare. Then over the great salt flats, gleaming in the distance like pure snow, and into Salt Lake City on May 23. Once again back into the lands of the Wild West, the convoy was met at Rock River, Wyoming, not by the usual police escort but instead by pistol firing cowboys who staged an ambush before the local sheriff arrived to lead the way into town.

During the course of the eastwerd route a message was received to say that, such had been the popularity of the campaign, the Canadian Government had requested that an extension be made to include some of the towns in that great country. Some lightning plans were made, to add 2,000 miles and another month to the itinerary. A survey of the proposed route was also made to eliminate time-wasting detours. Before crossing the border, two days were spent at Detroit, home of the American motor industry, and it would be here that the eyes peering under the bonnets would be sharpest, and the questions most informed.

The Canadian section of the tour took in 26 towns and cities in the eastern Provinces including Montreal, Ottawa, Toronto and, most appropriately of all, London, the city at which perhaps the most generous of all welcomes was offered. As if in token repayment nearly 3,000 rides were given there by RT 2776 in the course of a single day, and the trips had to continue until the early hours in order to satisfy the demand !

As neither the United States nor Canada were signatories to the international agreement of 1926 that allowed foreign visitors to use their own registration plates whilst visiting other countries, the buses collected additional insignia whilst on their travels. First of all,

Wild country and a precipitous drop in the Sierra Nevada Mountains.
[California Public Works Dept.

Fortunately a rough high-level track enabled this railway to be crossed without much difficulty when the bridge (left) on the normal route to Montreal proved too low.

[E. T. Bonny collection]

besides their own United Kingdom registrations, there had to be added GB plates. Then, once in the United States, their registrations were allocated by New York, and licence plates duly added. Canada was to add two more, from Ontario and Quebec.

Twelve thousand miles later, the journey was to end as it had begun, on the New York docks alongside 'Parthia'. Sometimes incredible scenes of welcome had become expected; indeed RT 2776 had carried some 57,000 passengers in the United States and another 25,000 in Canada, and over 132,000 people had visited the exhibition in RTL 1307. What must have come as something of a shock though was the crowd that awaited their return on Horse Guards Parade, where an official welcome back to London was given. After fifteen minutes of speech making, RT 2776 went straight back into service on route 11, still sporting the array of flags that had been worn for the reception.

In preparing this account the greatest use has been made of contemporary newspaper reports and magazine articles. It is a pity that a full list of the names of those who did the driving, conducting and maintenance work does not seem to have been published, for whilst one would not denigrate the efforts of those who undertook the planning of the venture, it was the men at the wheel and on the platform that created the tremendous success that it was. Perhaps, if they should happen to read this somewhat incomplete account (for there must have been many interesting features of daily routine that went unrecorded) then they would accept it as being a tribute to their work.

Since 1952 other London buses have been on visits to places far from their home city— San Francisco has been able to sample the qualities of a Routemaster, and a similar vehicle has graced the streets of Tokyo, albeit at the end of a long sea voyage. The memory of the three London buses has lasted long in North America: one similar machine has found its way into the private service of a New York restaurateur, whilst in Canada two companies have been formed to exploit their tourist attractions. One such company, London Omnibus Tours, of Victoria, British Columbia, owns two former Trent vehicles, whilst the other, Double Deck Tours Limited, owns three vehicles, which are believed to be of London's RTW-class. Both concerns have expressed a wish to buy more British double-deckers.

It will be a long time before the achieve-

The three 'musketeers' back in familiar surroundings at Chiswick Works.

[*Westinghouse Brake & Signal Co.*]

ments of 1952 can be bettered, taking technological advantages into account. Perhaps, in several decades, London Transport might be persuaded to exhibit a trio of the grandchildren of the Routemaster in a British Trade Week at Buenos Aires, and decide that much valuable publicity would be gained by sending them by road all the way! After passing through the Channel Tunnel, the congestion of European roads could soon be left behind in favour of the vast open spaces of the Soviet Union. The dam across the Bering Straits would no doubt be provided with a roadway, and having reached Alaska, it would only mean following the Alaskan and Pan American Highways to the ultimate destination. But that will be a long time coming; meanwhile, RT 2775 and 2776 are continuing to serve Londoners, but RTL 1307 has gone abroad again—this time permanently—and for the past two years has been resident in Ceylon.

Hay-Day for a Guy

by G. R. Mills

Above. Hay-day for CDR679, originally Plymouth CT No. 249, and now a useful utility at Wormingford, Essex.

Below. Another Guy—JNU679—ex-Midland General, as a shownam's utility at Cambridge autumn fair.

[G. R. Mills

AS with any motor vehicle, the fate of a bus after it has served its useful life as a public service vehicle can vary considerably, depending largely on its condition; it may go directly to a scrapyard and disappear for ever swallowed up by heaps of rusting metal. Fortunately, however, many escape immediate extinction by entering service with a variety of non-psv operators, principally as staff buses with building and civil engineering contractors, whilst many hundreds can be seen in use with showmen. Housing estate developments have encouraged the use of a vast number of former psvs as mobile grocery shops; similarly others can be found as mobile fish and chips bars and showrooms displaying anything from lampshades or furniture to paints or even machines, like lathes, installed is such a way that they can be demonstrated in working condition. The war years produced psv conversions to ambulances, firetenders and "clubmobiles", whilst the poor avaliability of commercial vehicles in post-war years encouraged a galaxy of reconstructions ranging from conventional wagons to specialised transporters, such as large milk floats, cattle trucks, horse boxes, pantechnicons and low-loaders. Bus operators throughout time have converted discarded buses to mobile breakdown cranes and towing wagons, while larger undertakings have produced tree-loppers, snow ploughs, vans for tickets, parcels or uniforms, shelter carriers, tankers, canteens and driving instruction units. Individual municipal departments have benefitted with such oddities as mobile libraries, toilets, clinics, invalid conveyances and vehicles constructed to carry road safety displays or as carnival floats.

Even dismembered psvs can provide useful items long after their passenger carrying life has expired, particularly with showmen and in agriculture where engines have been utilised as worthwhile stationary power plants and many chassis have provided sound bases for trailers; one less fortunate frame, however, serves as a bridge over a stream in Suffolk! Old bus and coach bodies have been used for

a wide variety of purposes on farms, including hay stores, hen houses, caravans, children's playrooms and even rare examples have been turned into a pump house, a dog kennel, a pigeon coot, a cycle shelter and potting sheds. An extremely unusual conversion exists on an Essex farm, well known in the motor-cycling world as the setting for regular scramble meetings where the normal marshall's tent has been replaced by a Bedford WTB, complete with a sentry box on the roof for the announcer.

Only a few acres of farmland away to the north is yet another former psv which, while staking no claims to uniquity, is certainly a rare bird in the district, for its very existence is due entirely to the presence of a bus enthusiast resident at the farm. The idea of buying a double-deck bus for conversion to a low-loading goods vehicle was the brain child of Mr. John P. Jackson, of Wormingford, who was both keen on saving hard work humping straw bales on to high trailers and in satisfying a desire to drive a double-deck bus. After studying the excellent specifications in back numbers of *The Commercial Motor*, it was decided that a vehicle such as a Guy 'Arab', preferably with a 6-cylinder Gardner or Meadows engine, would be excellent; alternatively a Bristol K6A or K6B was just as desirable as they offered a reasonable ground clearance.

The seemingly easy task of finding a suitable machine was put around to local enthusiasts, the requirements being that the bus must have sound mechanical units but the state of the bodywork was immaterial—it could even possess the prize waist-line sag of the district—provided the purchase price was well below £100. Numerous ideas came forth, but most suggested operators who had laid up good runners through lack of body repair facilities. A dealer's name was mentioned, but was squashed when it was learned from several sources that transactions were conducted in an appalling way. A scrap metal merchant was visited, when it was learned that he had purchased three municipal Bristol K6As for a total of £202 10s., but as the selling price was £120 each, and thus more than the budgeted figure, it was agreed that such excessive profiteering by a scrapman could not be encouraged!

Independent operators proved most helpful, measuring ground clearances, discussing bodywork fixings and suggesting additional rigs and bracing for such a conversion—but alas, no bus! On inspecting two Guy 5LWs that seemed to be asking to be put to work again, it was found that a cylinder head had been removed from one, whilst the other was without such essential items as a water pump and battery, and was reduced to single, bald tyres at the rear. Further enquiries revealed more Guys with curious 'S' certificates and, again, too highly valued by their previous owner. Enquiries were made about other vehicles known to be disused, only to find that the ever popular scrapmen or dealers had preceded us and thwarted us yet again.

Several East Anglian psv men had recommended Thurgoods of Ware as being very reasonable dealers, always helpful and obliging. It was then recalled that a Guy

CDR679 as spotted by the author in the Ware yard of W. L. Thurgood (Coachbuilders) Ltd. at Easter 1963.

[G. R. Mills]

'decker had been seen at their premises a few months previously—at Easter, 1963—and we wondered if it was perhaps still for sale. True to form, Mr. A. C. Ledger, of Thurgoods, was interested in our project and most keen to sell the vehicle, which was obtained for a price reasonably below that painted on the windscreen.

So, on June 19, 1963, CDR 679 left Ware, Hertfordshire, and climbed the steep Widbury Hill bound for Wormingford, Essex. It covered the journey of over 60 miles—passing through Bishops Stortford, Great Dunmow, Braintree and Coggeshall (a section of road served by ex-Moore Bros. Guy 5LW buses)—

without any trouble whatsoever, on what was to be its last day out on the public highway as an omnibus. Following behind was a blue Ford Thames 5 cwt. van with almost every conceivable roadside repair tool on board, but the faithful Guy/Gardner combination proved again its complete reliability, even after a six-month rest, and maintained steady progress throughout the trip. The fact that the 'old banger' had a current psv licence with a certificate of fitness that did not expire until January 1964—a full six months away—further ensured that the CDR 679 was in excellent fettle officially. We were thus convinced it was a bargain.

The bus, an 'Arab' II (5LW), chassis FD 26342, was first registered on October 27, 1943, being one of the large batch of Guys delivered to the City of Plymouth Transport Department to replace the many blitzed and damaged Leylands. The austerity bodywork by Charles H. Roe was of lowbridge type and provided accommodation for 27 persons on the upper deck and 28 in the lower saloon on wooden seats; these were replaced by upholstered leatherette covered ones when the vehicle was extensively rebuilt by Bond, of Wythenshawe, late in 1952. Redelivery for service was made in January 1953, with non-utility roof domes and rubber-mounted Solovent windows including two new rear upper deck windows in the emergency exit which had previously been a complete metal sheet. Other modifications included removal of the Roe trade rail, deeper skirt panels with the nearside section shaped to meet the front mudguard, and small refinements such as the building of an illuminated rear registration plate into the bottom offside corner in place of the original one painted on the top left-hand corner of the platform window which relied on light drifting back from the lower saloon!

During her career, CDR 679 has had four owners and has been graced with five different colour schemes. She was delivered to Plymouth in wartime grey, but was repainted in standard red and cream livery in 1946. In her post-rehabilitation days with PCT, she was usually to be seen at work on the circular services 36/37 serving City Centre-Peverell Beacon Park-Milehouse-City Centre, and was finally withdrawn on the last day of May 1957. The following month she moved to her next owners, Grenville Motors, of Church Street, Camborne in Cornwall, who first taxed her on July 2 and transformed her coat to cream and black. For four years she was a regular sight on Grenville's half-hourly local service between Cambourne and Troon, although she was also seen on the Cambourne-Penzance service on occasions; however, CDR 679 was finally displaced by a Leyland PD1, also ex-PCT, in December, 1961; an ex-PCT Leyland TD5 had been on the route before the Guy. Later that month, CDR 679 made the long pilgrimage to the Ware premises of W. L. Thurgood (Coachbuilders) Ltd.

Her sojourn at Ware was short for, by February 1962, she was again at work—this time on a schools contract in the Harlow New Town area in the employ of Harlohire Ltd., of Epping in Essex, who repainted her a pale green relieved with mid-green wings and roof. As with all Harlohire vehicles, their owner, Capt. Plunkett, is a great believer in the efficiency of bold lettering and the Guy 'decker received dark green block letters on the front destination screen glass and the rear lower panel declaring the fleet name LANDLINER together with Harlow Essex and the telephone number.

CDR 679's final conversion began soon after the Guy had arrived at Wormingford, but not before a party of enthusiasts had been given a tour of the 850-acre farm, where the old bus was to be put to work. Part of the acreage includes a section of a wartime airfield, where the vehicle looked somewhat out of place when negotiating the perimiter tracks and passing the remains of former bomb bays. The T-type hangars no longer exist, having been dismantled by contractors in recent years, but two small light aircraft—owned by a local business man—still use the remaining runways. Unfortunately, no direct access then existed between the two areas, so the Guy could not be photographed with the planes; the long way round by public road was also impracticable, as the vehicle has been limited since July 10, 1963, by a special 'Exempt from Duty' licence to six miles per week on the public highways. At the time of the special run for the lads, the Guy was not taxed or insured with its new owner, but that did not prevent driver John Jackson from putting the bus through its paces—to the delight of those on board—along the network of private roadways.

HAY-DAY FOR A GUY

In its heyday as a bus—CDR679 at Camborne, Cornwall (*above*) in 1960 on the half-hourly service to Troon operated by Grenville Motors, and (*right*) at the "Swan", Stanway, Colchester, in 1962 after acquisition as the "Landliner" in the fleet of Harlohire, Epping, Essex.

[R. C. Sambourne: G. R. Mills]

Left. This view of CDR679, after acquisition for conversion, shows the rear destination box and registration plate that turned out to be useful items in the reconstructed vehicle.

[G. R. Mills

On Wormingford Airfield in June 1963 after John Jackson's team had fitted a variety of destination blinds.

[G. R. Mills]

The first modification the bus received was a minor, but nevertheless conspicuous, one as the front destination screen glass was treated with paint stripper, revealing the three-part Plymouth-style layout of blind displays. The box was then fitted with a Swindon blind from a utility Daimler 'decker in the terminus section; a Devon General/Grey Cars blind, formerly fitted in an AEC 'Regal' III (via Greenslades), was found suitable for the via point section, while a hand-made set of route numbers was kindly supplied by a municipal transport traffic clerk who had painted them before joining the undertaking. The results were most pleasing and, at the time, it was suggested that the destination box should be fitted with a special blind, once the vehicle was engaged on its proposed new role, listing names such as Haystack—which actually appears on ENOC buses operating in the Canvey Island area—Church Hall, Hall Farm, Airfield, Sandy Hill, Fir Tree Hill, Lakes, etc., but unfortunately the conversion was more useful when devoid of such a refinement!

By nightfall on June 30, the Guy stood silent and forlorn in a steel-framed hay barn; she had been deprived of all her seats, interior light fittings, bells, buzzers, lower deck windows and destination equipment. The body panels had been carefully cut above the upper floor level over the canopy, and, over the remainder of the side structure, they were cut below the upper deck floor level.

The principal timber members were, surprisingly, rot free and cutting through them caused far more perspiration than had been expected from first appearances; comments from various sources, regarding the ugly rehashed body design and generally neglected look caused through lack of use, suggested a far simpler operation.

The next morning the stair fixings were severed and the staircase removed en-bloc—revealing a number of PCT and Grenville Motors tickets tucked away in the conductor's waybill rack—together with all the plastic-covered hand rails, which when piled on the ground looked somewhat like the framework for a fantastic modern art statuette. A large timber bearer, of similar section to a railway sleeper, was then lashed under the middle of the upper deck and attached to a chain pulley block secured to the roof trusses of the barn. With the help of a tractor fitted with a scoop, the rear end portion of the bus was prised apart at the saw cut and the pulley block used to lift the top deck clear; but as the upper body hung in space, the vehicle refused to start and had to be towed clear by another tractor. The fault was found to be a jammed electrical contact breaker, which had apparently given trouble before as the casing was dented.

The conversion sped on; the side panels were cut down to just below wheel-arch level, and the platform became a sub-floor for new timber planking on studwork to make the height of the floor uniform throughout. The space between the new floor and platform has provided a useful compartment for pitch forks, lifting jack, wheel brace, starting handle and other essential items of equipment. John Jackson was very lucky on two points with the reconstruction of the bus rear; first, the rear registration plate fitted exactly in the space provided by the two floor levels and was attached to the old and new timberwork; secondly, the old two-aperture rear route number box—a Plymouth feature but used by Grenville Motors for their initials—was, by chance, ideal as a lift-up flap on the locker.

Several test runs were made when the conversion had been completed, and the well known vibrations of the throaty gurgling Gardner 5LW in an 'Arab' chassis, coupled with the fact that the terrain over which the vehicle was tested was not as kind as the public highway, were soon apparent. A fair amount of cab structure sway was felt, which was not surprising after the loss of rigidity following the removal of nearly $2\frac{3}{4}$ tons of bodywork! The addition of a length of steel angle across the radiator with $1\frac{1}{4}$ in. bore pipe stays strutted back to the dash and canopy considerably reduced the cab movement; these additions have given the external appearance a very rugged look, not unlike that associated with army vehicles.

During August that year, CDR 679 was to be seen carting hundreds of straw bales from the adjacent Church Hall farm fields across the main road from Colchester to Sudbury and Bury St. Edmunds—served by H. C. Chambers & Son, of Bures, with Guy/Gardner 5LWs—into the main Wormingford Hall farm and travelling for nearly a mile on private metalled roads to the barn where the decapitation began. The Guy's top deck then rested on concrete blocks alongside this building, awaiting a purchaser, but no offers for an 'ideal' dog kennel, garden shed or chalet, or even a garage for a mini-minor or sports car were forthcoming. From personal experience of the warmth of the interior—access could only be gained through the destination box!—in summer months, the upper deck would have provided an excellent hot house for pot plants. As none of the sales ideas reached fruition, the aluminium panels proved very useful in the construction of a cab for a beautifully restored Morris Commercial 4 × 4—formerly a 'member' of H. M. Forces but now inappropriately registered NVW906C—while the communication tunnel between this cab and the truck has a patent sliding ventilator!

By late September '63, harvesting having finished, the fields were alive with tractors engaged on drilling operations, and the former bus was again on the scene laden with piles of hundred-weight bags of artificial fertiliser. Only after heavy thundery autumn showers did some of the steeper unmade farm tracks prove difficult for the Guy to navigate, and wheel spin occurred when the somewhat smooth rear tyres failed to grip.

Since the initial conversion, various odd fitments and modifications have been found necessary; these have included a pair of spikes

One-man-operation in the hay barn at Wormingford Hall Farm on July 1, 1963; the lower half of CDR679 being towed forward as the upper part remained suspended from the barn roof.

[J. P. Jackson]

on the extreme front corners of the canopy to secure bales being carted, and the filling of the slightly sunken gangway with boards to give a completely level floor so as to facilitate the use of sack barrows. Much of the internal moulding and polished wood sills of the original body has been used to trim the cut down sides to give a safe working edge, and the cut-away section of the offside upper deck lowbridge sunken gangway on the canopy has been neatly boxed in with timber.

A handbook, kindly loaned by a municipal transport undertaking which no longer operates Guys nor 5LW-powered buses, enabled an extensive routine maintenance

The last straw—three licensed p.s.v. drivers and three licensed conductors—one with a former Midland 'Red' Setright ticket machine—took part on this overstaffed (?) journey round the farm by P.S.V. Circle members in August 1964.
[T. M. Smith]

check to be made. Then advantage was taken of the mild weather in late October and early November to roughen up the Harlohire green paint with a rotary wire brush, after which two complete coats of pale yellow undercoat were applied. But as this colour clashed with the bright red of the lining out, an eye catching buttercup yellow was chosen instead, and the wings, wheels and mouldings were carefully picked out in GPO red. The finished result, at first glance, is somewhat similar to the smart open toppers used by the East Kent Road Car Co. on the sea front services at Herne Bay and Margate, although the structural alterations have been considerably greater than to the East Kent vehicles.

The completed project is a credit to the ingenuity of John P. Jackson, who undertook, and successfully overcame, several difficult tasks during the conversion. The rebuilt and repainted Guy was completed by January 1964, and from that time CDR 679 has been ready for every busy farming season. In addition to the CDR registration mark identifying the original home, the PCT fleet number—249—is still carried on the varnished bulkhead woodwork, whilst the demanding cautionary notice that Smoking is Prohibited still very much applies. In retrospect, although there is a vast contrast between CDR 679's municipal service in Devon and her present role in Essex, it must be recalled that Plymouth Corporation buses have a reputation for long life in others' hands; former PCT vehicles have at various times seen service with Eastern Counties, United Counties, Thames Valley, Crosville, Eastern National, United Auto, Bristol Tramways, and have strayed over the Scottish border to W. Alexander, Scottish Omnibuses and Western SMT; the Welsh hills have seen them in the fleets of United Welsh, Red & White and Jones of Aberbeeg, and they have even turned up with Jersey MT and the Northern Ireland Road Transport Board!

In conclusion, thanks must be extended to Mr. R. C. Sambourne, of Swilly, Plymouth, for providing details of CDR 679's life in Devon and Cornwall.

Co-ordination

by Charles F. Klapper

DURING 1966 co-ordination of bus operations with the activities of other forms of transport has again been a live issue. For many years there has been a vociferous demand for better rail and bus connections and this has been stimulated by the railway closures following Lord Beeching's famous report on the reshaping of the railway system.

Those who wish to travel by public transport from start to destination find, often enough, that the initial hurdle, in planning a long journey, is to discover the bus operator at the far end, whether the trunk part of the journey is to be made by train or coach. Some bus offices, such as those of Ribble Motor Services, are better provided with practical travel information than the average railway inquiry bureau; in the Metropolis, London Coastal Coaches is resourceful in helping the traveller on his way. Many ordinary travel bureaux would sooner help a passenger to fly to Timbuctu than put him or her on the right bus for Todmorden. Especially if they elect to travel by train travellers may find a considerable gap between railway station and bus station in the town where the interchange is being made or that at the station where they intend to alight the bus has gone a few minutes before their arrival. The defects of geography may have to be borne; it is hard to put up with what look like defects of timetabling.

Yet these lapses may be just as hard to remedy; the bus can wait only a limited time for a late-running train; the bus has a more important connection to make with another bus service or different railway service further along its route; in any event it is a record if at an interchange point as many as $4\frac{1}{2}$ per cent of the bus passengers are interested in transferring to or from a train, so that dissatisfaction for the majority arises from giving a service to the few.

In some other countries, to assist in connections being made and held for through rail passengers, connecting buses are firmly labelled with their purpose. Railway-connecting buses in France appear usually with the names of places served on a band along the side, ending with the unmistakable *SNCF Correspondances*. In Belgium, although as in France operated by contractors to the railway, the railway-associated buses are in a uniform blue livery chosen by Belgian Railways and bearing the prominent B totem of the railway undertaking. In Germany, Austria, Denmark and other countries they are plainly the property of the railway and operating services well-publicised by the national railway undertaking in its timetable. It is an interesting theory to pursue as to what benefits might accrue in Britain if railway substitution buses were in the standard BR blue livery and carried British Railways as a fleetname, plus "connecting with trains at Banbury" or whatever the appropriate destination blind might be.

Whatever the merits of railway connections, the discipline of bus link-ups is paramount. Several provincial bus undertakings owe their rapid build up in the early days of the motor bus to the judiciously contrived network of cross-connecting routes provided across their territories at regular intervals throughout the day. Sometimes these may involve some slow running or waiting to keep a scheduled connection, but the value of the facilities to travellers more than justifies this. In the 62 miles from Gravesend to Brighton, the Maidstone & District and Southdown joint route made, before recent railway closures, no fewer than 25 sets of connections (at 17 places to other bus services and at eight to railway services) and thousands of other instances could be quoted.

The story of road and rail co-ordination goes back to the earliest days of railways—the Liverpool & Manchester Railway not only carried road vehicles on its flat wagons but instigated bus connections with its trains. Another early example of rail and road co-operation was at the opening of the London & Birmingham Railway on April 9, 1838 from Euston to Denbigh Hall, just north of Bletchley. Stage coaches provided the 35-mile connection by road to Rugby, where the train journey was resumed. The Great Western Railway offered financial induce-

Railway Replacement A former UAS group Bristol handed over to N. Fox who operates in place of the Border Counties Service between Bellingham, Kielder and Steeles Road station on the Waverley route to Edinburgh.

Temporary Train Replacement. A London Transport RF type AEC replaces trains from Elmers End during Southern Region week-end engineering works.

Use of Former Railway Property (left). The Ribble bus station at Southport was formerly the Lord Street station of the Cheshire Lines Committee.

Railway Connections (below). Crosville Bristols at Towyn station; the integral bus on the right is on the service to Dolgellau via Dolgoch Falls and Talyllyn Lake.

[C. F. Klapper]

ment to bus proprietors to go out into the fields to pick up passengers at Paddington Station, then situated on the fringes of the built-up area. Occasionally a railway company invested in its own buses in order to feed traffic on to its own line and the Metropolitan Railway ran three-horse buses designed to link its line on the northern side of the Circle with the bright lights of Oxford Circus and the West End.

The real drive came with the motor bus, when in 1903 two Milnes-Daimler buses were obtained to run in conjunction with the Lynton & Barnstaple Railway to link its Blackmoor Gate station with Ilfracombe. Hostility by Ilfracombe horse coach proprietors defeated the project within hours rather than days and the Great Western Railway was persuaded to take over the buses. So on August 17, 1903, they were put to work from Helston to the Lizard in lieu of building a costly but doubtless unremunerative light railway. It was not long before other railways followed this course and the GWR soon had one of the largest provincial bus fleets. Later the process worked in reverse and the Lampeter-Aberayron bus route was replaced by a light railway. In the case of some railways, such as the Great Eastern, buses were from about 1912 handed to bus owning firms. In 1913 the GER petrol buses around Chelmsford, including some remarkably heavy home-built vehicles, were replaced by the steam buses of the National Steam Car Co. Ltd; from this small beginning sprang the Essex and West Country activities of the various National companies of the present day.

The four grouped railway companies obtained road transport powers in 1928 and at first it appeared that they were going to offer sharp competition with the existing large bus companies, which had already hedged in their operating areas by mutual agreement so that they did not indulge in useless competition. Competition in transport is especially damaging because the seat that goes empty at this instant moment can never be of use to anybody—it is lost for ever to the passengers or as a means of revenue to the operator of the service. At the insistance of the Royal Commission on Transport which was then in session the expected competition was exchanged for co-operation.

The railways invested in up to 49 per cent of the ordinary shares in most of the principal operating companies and began, through working committees, to arrange for connecting services, inter-available return tickets, mutual publicity in timetables, bus services to cover interruptions to railway services, coach tours to supplement railway excursions and many other facilities. There were also sharing arrangements in the outer zone and interurban business of four municipal undertakings—Halifax, Huddersfield, Sheffield, and Todmorden — all in West Riding. These arrangements continue today between the railways and the nationally-owned bus companies administered through the Transport Holding Company and also with the companies of the British Electric Traction Group (in many of which the railway shares, totalling up to 49 per cent are now the property of the nation's Transport Holding Company); they provide a framework of train and bus coordination that has been strengthened by the arrangements made to cover withdrawn train services and can perhaps be developed further to provide a still more useful network of travel facilities for those who desire to rely on public transport rather than the motor car.

Co-operation between bus operators took place within two years of the new vehicle being started in London, plying for hire on its way, by George Shillibeer, on July 4, 1829. Competition had become extremely fierce and

Railway Interest. Through joint committees the railways have a 50 per cent interest in four municipalities. This Huddersfield trolley is at Crosland Hill on the Birkby service during the last week of operation; railway and municipal services are now all operated with motor buses.

operations were rationalised by the formation of associations, each of which managed a route and apportioned the buses (or "times", which became a valuable commodity) upon it. In various towns, such as Bristol, Brighton, Liverpool or Manchester, mergers of groups of proprietors into one entity took place with the same object; the formation of the London General Omnibus Co. Ltd. is perhaps the best known, because at first, in 1856, it was conducted under French auspices and was known as the Compagnie Générale des Omnibus de Londres. Several of the area agreement bus companies of today, such as East Kent and East Yorkshire, are the result of several companies being brought together to provide greater flexibility of vehicle, engineering or financial resources.

The joint service operated by agreement between two or more operators evolved in various circumstances. The earliest example among electric tramway operators appears to date from 1904 on the Barking Road in the London area. It was inaugurated by East Ham and West Ham Corporations; the distinguishing feature was the simplicity of the accounting arrangements. In each municipality's area, the owning undertaking provided the conductor; although the drivers worked through, the conductor left the car at the borough boundary with cash bag and tickets, to be replaced by his opposite number from the other owning authority. Even today the stark simplicity of separate ticket issues is preferred at certain borough boundaries to the refinements of modern accounting systems.

Pooling of receipts on joint services can take many forms and often is on a different basis for each of several joint routes set up by the same operators in order to meet different traffic conditions. As long ago as 1937 Ribble Motor Services had more than a score of different schemes with the companies and municipalities with which it came into contact for joint operation or for passing through "foreign" territory. From the public viewpoint, joint timetables, the same service number, inter-available return tickets, mutual acceptance of season tickets and parcels, and the same conditions for carriage of dogs and parcels, are vital on joint services.

The coaching pools are responsible for very large passenger movements. The Limited Stop Pool originated in a Northern General service from Newcastle to Liverpool begun in 1928; in 1929 West Yorkshire, Yorkshire Woollen District and North Western Road Car joined in; Lancashire United joined in 1932; and United Automobile Services in 1934. Yorkshire Services began as a joint arrangement between West Yorkshire, Yorkshire Traction and Yorkshire Woollen District in preference to competing on Harrogate-London and Harrogate-Birmingham services; East Yorkshire joined the pool in 1931. In 1934 a Yorkshire-Blackpool pool was set up.

In the same year the Associated Motorways pool based on Cheltenham came into full operation. The station of Black & White Motorways (itself a Tilling-BET joint concern with shareholdings by Bristol, City of Oxford and Midland 'Red') was used as headquarters and the timetable of interchanges of service from all points of the compass revolves around the meal stops on the through services. The original participants were Black & White, Greyhound (later absorbed by Bristol), Red & White, Royal Blue, Midland 'Red' and United Counties. Of these Royal Blue was independent and shortly was acquired by Western and Southern National; Red & White was purchased by the Tilling group in 1948. The railway interest in Associated Motorways amounted at first to only 35 per cent, whereas it was 49 per cent in the Yorkshire Services pool.

Violent competition was sometimes avoided by allocation of territories. In the autumn of 1914 a statesmanlike agreement was effected

Railway Participation. Sheffield Corporation operates its A fleet—of which this Daimler 'Freeline' at the builder's works is an example—in the inner area; the B area is shared with British Railways, and the C fleet operates on behalf of the railways.

[T. W. Moore

CO-ORDINATION

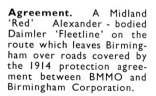

Agreement. A Midland 'Red' Alexander - bodied Daimler 'Fleetline' on the route which leaves Birmingham over roads covered by the 1914 protection agreement between BMMO and Birmingham Corporation.

between the Birmingham & Midland Motor Omnibus Co. Ltd., (Midland 'Red') which had been established in the running of local bus services in the city 10 years previously, and Birmingham Corporation, the tramway undertaking of which had begun to add feeder bus services to its resources, beginning with one from Selly Oak to Rednal, during the previous year. The BMMO handed to the Corporation the fleet of 30 buses it operated in the city and a garage; the company agreed to restrict its activities to services running outside the city boundary, a protective fare in excess of the appropriate municipal tram and bus fare being charged in respect of passengers both picked up and set down inside the municiple boundary.

This pattern of division of function, adapted in respect of other municipal systems and other bus operators to varying needs, although not of universal application, has in fact, been one of the mainsprings of the development of the widespread network of interurban bus services in Great Britain, which has the benefit of a greater number of interurban buses and coaches proportionately to its area than any other country. The arrangements ensure that the municipal operator has a near-monopoly for development of the entire area, so that it can take the rough with the smooth and sustain necessary but possibly unremunerative routes from the profitable ones; the interurban company operator secures a higher fare, and, because of spending less time on local picking up and setting down, is able to offer a higher schedule speed; the third and most important party to the transaction, the passenger, has the benefit of a faster transit to the city boundary and, above all, can board an outward-bound bus at the city terminus without being crowded off by local passengers.

These matters are so important to public service that it is certain in Birmingham that even if the two undertakings were the property of the same authority they would have to be continued for the general good. Moreover, flexibility has enabled the town and country undertakings to cope with changes in the municipal boundary and to deal with the situation concerning fringe services to estates just beyond the boundary so that in later years BMMO has by agreement worked what were virtually city services and were able to do it with economy and to the public benefit.

The system here of getting country passengers direct without change to city centres may be contrasted with the insistence, in numerous cases on the Continent, upon passengers changing vehicles at the municipal boundary (or even more arbitrarily at a suburban tram or trolleybus terminus) or at some out-of-the-way station on a local railway in order to make a comparitively short journey into a city centre. In Great Britain passengers resent having to change and a 1962-63 experiment in Glasgow to persuade residents on the Castlemilk estate that their best route into the city centre was by bus to King's Park station and thence $3\frac{3}{4}$ miles by the Blue Train electric service to Glasgow Central was a failure. The three minutes allowed for changing from bus to train gave a direct bus too big a start on those who completed their journey by train.

The Traffic Commissioners appointed under the Road Traffic Act, 1930, either suggested or enforced rationalisation of services, joint timetables (and even introduced rather odd operating methods such as the journeys which were operated in alternate weeks between Fort William and Inverness by MacBrayne and McRae & Dick) and equalised fares; often the upshot was the purchase of one operator by another, resulting in uniformity and regularity of service without cut-throat competition, fare-cutting or dangerous situations such as rival vehicles converging on the same queue of passengers simultaneously.

One example of co-ordination of two operators to which experience of the 1930 Act quickly gave rise was that of East Yorkshire Motor Services and Kingston upon Hull Corporation Transport, devised in 1933, as a result of which there has been a combined effort since July 29 1934 to avoid overlapping and to spread bus mileage as efficiently as possible over the outer areas of the city. There are three zones: Area A represents roughly old city limits; Area B extends well beyond the city boundary to Hessle, Willerby, Wawne, Ganstead, Preston, Hedon and Paull; Area C is the territory solely covered by the East Yorkshire Company services. In the A Area the receipts from all A Area passengers go to the Corporation; in the B Area revenue and car-mileage are equally divided. In the C Area they are of course solely attributable to East Yorkshire. A practical feature of the implementation of the Hull scheme was the erection of posts showing the boundaries of the A and B zones as a reminder to conductors. The scheme has worked harmoniously under a committee of equal numbers of company and municipal representatives and in 1935 the two undertakings jointly purchased the bus business of Sharpe's Motors Ltd. The public has had a better service at lower fares than could otherwise have been possible.

From the time of the establishment of the Traffic Commissioners onwards large numbers of mutual help schemes or of small concerns being taken over by larger on a 21 year lease or similar basis were arranged, to secure the economies of larger scale operation. Municipal undertakings at such diverse places as Ayr, Perth, Dover and Worcester were thus taken over by a large company neighbour; the neighbouring Burnley, Colne & Nelson undertakings were merged for operational and administrative purposes; and the Peterborough Electric Traction Co. Ltd., Ortona Motor Co. Ltd (Cambridge), and Eastern Counties Road Car Co. Ltd (Ipswich) were merged with the Norfolk and Suffolk areas of United Automobile Services Ltd. to form the Eastern Counties Omnibus Co. Ltd., based on Norwich.

During the past quarter of a century a number of urban and interurban operators have arranged complete sharing arrangements, pooling receipts and mileage worked on an agreed basis. One of the first agreements was between Brighton Corporation and Brighton Hove & District Omnibus Co. Ltd., however, where parts of a conurbation were served by the company and other sectors by the municipality. The pool ran satisfactorily for 21 years on a $27\frac{1}{2}$ per cent-$72\frac{1}{2}$ per cent

Forced Co-ordination. The German method of enforcing tram/bus co-ordination, not without considerable public inconvenience—a scene in south Augsburg where city trams terminate in a loop to connect with outer suburban buses.
[C. F. Klapper]

CO-ORDINATION

basis and then a new agreement brought in Southdown Motor Services Ltd. to cover buses or interurban routes and in fringe suburbs bordering the boundary between the operators. The share is now on a 20½ (Corporation)-50½ (BH&D)-29 (Southdown) basis and the territory is served with great efficiency, routes having been rearranged to deploy fewer buses to greater advantage. Not only fares but conditions (such as carriage of dogs) are standardised; the same ticket machine is used and there is a jointly issued timetable book for the area from Rottingdean through Brighton and Hove to Shoreham.

Pooling (*above*). In Portsmouth mileage and receipts are shared in agreed proportions—a Southdown bus for Warsash and a Corporation vehicle to Paulsgrove.

Tripartite Co-ordination (*left*). Brighton and Hove provide the classic study in co-ordination—for 21 years Brighton Corporation and BH&D pooled receipts and mileage; then the local Southdown interests were added to the facilities of Brighton Area Transport with complete co-operation between the three operators. This information display is in a shelter at Brighton railway station.
[*C. F. Klapper*

Many similar arrangements are in force—Portsmouth Corporation and Southdown, Plymouth Corporation and Western National, Exeter Corporation and Devon General, Gloucester Corporation and Bristol Tramways & Carriage Co. Ltd., Southend-on-Sea Corporation and Eastern National are examples. At Bristol the Corporation purchased the City business from the company and handed it to Bristol Omnibus Services (as it is now known) to operate; in this case city buses are lent to country routes at the weekends and country buses assist in handling the city peak traffic morning and evening. This is a typical example of co-ordinated working benefiting both operator and the bus-riding public. The conurbation studies which have been carried out in the Glasgow, Merseyside, Manchester (the South East Lancashire and North East Cheshire, or SELNEC scheme), West Riding, Cardiff, Bristol and Newcastle upon Tyne areas may produce even more interesting results to the student of street transport.

The Ramblings of an Enthusiast

by Oscar Alison

I SUPPOSE I am one of the lucky ones whose principal hobby has now turned into a very worthwhile and absorbing livelihood. To a certain extent it is still a hobby as one can always learn from what the other fellow does and that means travelling and looking at other bus undertakings.

My early efforts at reading were apparently assisted by a close study of the then current issues of *Tramway and Railway World* and the *Electric Tram, Bus and Coach Journal* and it follows quite naturally that I took a keen interest in holiday and other journeys.

One of my earliest memories is of happy afternoons spent on the erstwhile Bents Green circular service 28 of Sheffield Corporation Motors as they were then known. This route, known locally as the 'Id dizzy' started from a most substantial stone-built shelter next to the Ecclesall tram terminus which it shared with the Dore route, proceeded along Ecclesall Road South and Bents Road for a brief halt at Bents Green, thence along Ringinglow Road to Ecclesall Road and back to the stone shelter. The fare for any distance on the loop was 1d, hence the title, and what better way of spending an hour or so at ½d half fare. The vehicles were generally Leyland 'Titans' and occasionally we were treated to one with open staircase body, all of course in the original Sheffield livery. The projection of Route 28 via Psalter Lane to a city terminus at Flat Street meant the end of these excursions but your scribe was by then fitted for more arduous journeys round the Outer and Inner Circles, even so they never conjured the same thrill.

Visits to grandparents took the family to St. Neots in Huntingdonshire, at that time served by the Midland section of Eastern National. The railway station was, and still is, some distance from the town and the absence of local bus services on a regular frequency left a gap in communications. This was filled by the proprietors of the Cross Keys Hotel in the Market Place, who operated a venerable single-deck bus, probably a Daimler, but I am not certain. The soundly built body had large windows and was finished in a canary yellow colour; the driver was a splendid little man, resplendent in leather gaiters. This vehicle would meet all trains and then pursue a tortuous course through the town, setting down the clients. Those wishing to catch a train made a booking at the Cross Keys and were told the time that they would be picked up as the bus made its rounds. Just before 1939 the original bus was replaced by a Dennis 'Ace' in a mustard livery, but the service was replaced during the war by a normal taxi service. In retrospect it would appear that the operation cut across all the requirements of road service licensing but nevertheless it filled a great need.

Seaside holidays at Flamborough meant travel on the wonderful vehicles of East Yorkshire Motor Services. With their primrose wheels and curtained windows, the single-deckers which ran to Flamborough Lighthouse or North Landing from Bridlington were the last word in comfort and efficiency. This was more than could be said for

Above. Sheffield 96—a 1934 Leyland 'Titan' TD3C with Craven 50-seat body—is an excellent example of the original livery and turnout of the Sheffield fleet.

[W. J. Haynes]

the vehicles of the White Bus Co. which ran a rival service for many years. Its fleet of secondhand vehicles, all of which seemed to be incomplete in one way or another, would have been manna for any enthusiast properly equipped with camera, notebook and tape recorder (if the latter had been available in those days). Occasionally East Yorkshire would send one of their ancient normal-control Dennis open-top double-deckers to Flamborough—these vehicles had a conductor's bell on the top deck front bulkhead whence there was no chance of seeing the state of affairs on the platform! Based on the four-ton chassis, these vehicles dated from the middle 'twenties and memories came flooding back when one was depicted in the second issue of *Buses Illustrated* in 1950.

At a later date, journeys between Leeds and Sheffield provided scope for "bussing". At that time there was no through operation but there was a choice of routes from Leeds. There was the conventional West Riding/Yorkshire Traction operation to Barnsley, via Tingley, Ardsley and Wakefield, invariably on a Leyland double-decker running cheek by jowl with Leeds trams on Dewsbury Road, Yorkshire Woollen District buses around Tingley and the red centre-entrance town service double-deckers of West Riding in the environs of Wakefield. Alternatively, there was the service by Burrows of Wombwell, which left Leeds via the Hunslet Road and proceeded via Woodlesford and Stanley to Wakefield, and thence via a devious route to Barnsley and on to Wombwell and Rawmarsh. Burrows sported a miscellaneous fleet, largely secondhand and at various times one could ride on an ex-Leeds AEC double-decker or even on an ex-London Transport STL. The Barnsley terminal was at the Yorkshire Traction bus station in the town centre where the red and cream Leylands were the mainstay of the operation. "Tracky" did have some pre-war Dennis single-deckers and obtained more after the war, and these vehicles mingled with those of Rotherham and Sheffield Corporations, Yorkshire Woollen District, West Riding, South Yorkshire, Rowe of Cudworth and other independents. The bus station was built in 1938 to replace street termini and at one time handled 15,000 departures per week. The journey to Sheffield was completed on a Tracky or Sheffield double-decker by way of the old Barnsley tram route to Worsborough, thence via Hoyland, Chapletown and Ecclesfield, entering Sheffield at Sheffield Lane Top and covering the tram route into the city. Nowadays the through journey takes two hours 22 minutes but with a 20 minute service into Sheffield the opportunity was generally taken to while away 40 minutes or so at Barnsley to watch activities in the bus station.

During the war years outings to the Ilkley area provided recreation and what better means of travel than by bus. Before the outbreak of war, Leeds Corporation ran services to both Otley and Burley-in-Wharfedale as successors to their early trolleybus ventures from the Guiseley tram terminus.

A 1935 Leyland 'Tiger' TS7 coach chassis of Samuel Ledgard as later fitted with the ECW 34-seat body from EYMS 340; note that all nearside windows can open.

[R. F. Mack

Former London Transport STL 1796 rests in Leeds bus station after a journey on the Burrows service from Rawmarsh; the addition of a cab door has little changed the 1936 Chiswick product.

[R. F. Mack

Left. An ex-South Wales AEC 'Regent' passing the Lawnswood tank barriers in 1940 on G. F. Tate's Ilkley service; the Brush body has an unusual nearside list.

Below. Though in Keighley-West Yorkshire livery, this 1932 'Titan' TD2 with Eastern Counties 53-seat body is typical of the West Yorkshire vehicles used on the Ilkley route.

[R. F. Mack

For a while the services were linked with the Dib Lane and Foundry Lane routes (67 and 68), thus replacing the former tramcar link from Roundhay and Harehills to Guiseley but the bus routes were pruned back to Guiseley in 1939, leaving the other operators with the freehold. The main protagonists were West Yorkshire and Samuel Ledgard, whilst a certain Mr. G. F. Tate also sent some venerable AEC vehicles to both Otley and Ilkley. This service yielded to Ledgard in 1943 and two AEC double-deckers changed hands as well. These had started life in South Wales and came north about 1940. West Yorkshire and Ledgard had a joint service to Otley/Ilkley from Cookridge Street, Leeds, via Lawnswood and Bramhope, whilst Ledgard ran their own service from King Street via Guiseley and Menston. Ledgards used early 'Titan' TD vehicles with Leyland bodies and the West Yorkshire offering was often a 'Titan' with Eastern Counties body, their standard type until the mid 'thirties. Both operators favoured low-height vehicles, though Ledgard began to use Roe-bodied full-height Daimler utilities on the Guiseley route from 1944. The resources of both undertakings were somewhat strained in those wartime years and mechanical unwillingness was not unknown.

Going west from West Riding brought one to the County Palatine and to celebrate the end of the war I ventured to Manchester. For speed (?) the journey began by train to Ashton taking, a trolleybus thence to the Cottonopolis. Manchester issued an all-day ticket at that time, 1s. or 1s. 6d. if I recall, and armed with one of these a great deal of tram, trolleybus and bus riding was possible. The various reds of the Lancashire fleets made a change from the municipal hues of Yorkshire, where only Huddersfield favoured red, and indeed the stately blue and cream of Rochdale buses looked quite out of place. Many of the Salford buses at this time were red and grey whilst some of that undertaking's trams appeared to be black—this I discovered was easily shifted by a passing coat sleeve or other portion of clothing! Another novelty of the area was the Crossley, which never seemed to make its mark east of the Pennines. Some of Leeds earliest deckers (1930) were of this marque, but only a handful, and the next delivery was not until 1948, but post war Manchester was rich in Crossley—oil and electric. The distincitve body style and livery of Manchester, with its curves, was a refreshing change from the usual straight lines of bus body builders and seemed to create a character so often lacking in bus fleets. Then, as now, Manchester was a mecca for the enthusiasts with the vehicles of about 10 municipal and half-a-dozen company operators to be seen on stage carriage operations.

A first post-war visit to London revealed how much havoc the war had wrought with the road transport fleet. My first journey was on a London Transport sixwheeler working on route 73—on receipt of the bell signal it appeared that the driver and chassis set off at a capital rate from Kings Cross; by the time they reached Euston the body got into the act and set off in hot pursuit! This phenomenon was not confined to this one vehicle, whilst others had obvious traces of the anti-blast net that had obscured the windows, white wing edges and white spots on the back for blackout conditions. The London scene had altered in other ways; the 'Utility' bus was there in strength and some of the Guys were finished in an all brown livery most unbecoming to London and the destination information had become severely restricted, indeed it never returned to pre-war standard as the ultimate reversion to full display omitted the rear number box. At the end of the war the trolleybus system presented much the same image as it had in its formative years, the borrowed plumes had returned to their proper homes and it was only in the Ilford and Barking area that the South African trolleybuses disturbed the near standard appearance of the fleet. There was certainly no suggestion that within 15 years the London trolleybus would be well on its way out.

East of London, the early 'fifties presented much of interest and one outing comes to mind. Starting from a Lyons teashop at Aldgate the eastward course was by way of Barking (and the South African trolleybuses) thence to Rainham where the red of Central Buses gave way to the Country Bus green. Route 371 from Rainham had been projected from Grays to Tilbury in January 1952 over the former Eastern National 31 route when the services of the two operators were integrated to overcome the barrier which had been erected at Grays War Memorial on 1st July 1933. It is interesting to recall that at the time of the integration, red buses of London Transport were used on the ex-Eastern National routes whilst ex-Eastern National Guy utility double-deckers and Bristol single-deckers carried the London Transport fleetname and sported red garage stencil letters. After an investigation of the London Transport garage and the former Eastern National premises in Argent Street, the journey was continued on route 2 towards Pitsea and Southend.

At Pitsea, Mr. J. W. Campbell operated 'Motor Services' from premises in Station Parade. In addition to his three regular stage services there were some contract operations, and notices on the reverse of his Bell Punch tickets proclaimed "Coaches for Private Hire, Removal and Haulage Contractors, Sand, Ballast, Bricks etc., Coal and Coke, Estimates Free". The depot yard, which contained several Albion single and double-deck buses and some ex-Leeds Corporation AEC single-deckers, showed no evidence of lorries of any kind; perhaps Mr. Campbell had another yard, though a quick inspection suggested that coal, coke, sand, ballast and bricks were not strangers to the interior of some of the single-deckers! Readers will probably know that Campbell succumbed to Eastern National and it is interesting to note that the Basildon New Town now stands across the Campbell territory; perhaps if the New Town had developed earlier, Mr. Campbell would still be in business with some lucrative town services, for Eastern National have built up a network of routes in the New Town. On from Pitsea the journey took us by way of Benfleet and Westcliff into Southend. The separate concerns of Benfleet & District and Canvey Island had been merged into the Westcliff fleet by the time of the journey but the ex-Birmingham utilities which had been freely purchased were still to be seen, though several were very much the worse for wear at the Benfleet garage. At this time Southend was in the middle of a change—the trolley-

One of Campbell's ex-Leeds Weymann-bodied 'Regals' of 1932.

[O. Alison

buses were under sentence of death and the oldest units had already succumbed. The Strachan-bodied AEC-EE vehicles shared the duties with some Sunbeams with Park Royal bodies, purchased from Wolverhampton. The traditional AEC-Weymann lowbridge combination in the bus fleet was giving way to the Massey-bodied Daimler or AEC marque and there was a distinctly sloping back atmosphere to the place. Exceptions to the rule included some Dodge single-deckers converted from RAF lorries and a couple of really venerable 'Regals' with English Electric bodies. The latter boasted Gardner 5-cylinder engines and sounded more like the standard Eastern National product which abounded everywhere. These vehicles had been part of the original fleet which had restarted municipal bus operations in Southend in July 1932 and were to be very much rebuilt to fit them for further service. For the few routes where full-height buses could pass, Southend had purchased a couple of 'Regents' from Leeds with the famous Roe bodies. One had already been cannibalised for engine spares. A tour of the town revealed a seafront boarding house with bed and breakfast for 5s. (but it was next to the gasworks) and took us along Thorpe Boulevard, where Southend's tramcars had found their way along a central reservation. This was now picturesquely overgrown and the size of some of the trees made it difficult to realise that even 3′ 6″ metals had passed this way. Arrival at Victoria Circus in the centre of Southend coincided with the contract invasion from Corringham refineries—a casual glance at the varied liveries created the atmosphere of Lower Mosley Street in Manchester or Wellington Street, Leeds. Many north country municipal, BET and Tilling operators were represented and the procession took about 15 minutes to pass—a fitting end to the day.

South-east from London lies the Isle of Thanet where East Kent reigns supreme. This operator managed without fleet numbers in the accepted sense but possessed an interesting fleet. I arrived just too late to see the Daimlers inherited from the Isle of Thanet Tramways but in plenty of time to sample the 'Titans' which had formed the bulk of the pre-war double-deck fleet. With new Park Royal or ECW bodies they compared well with a batch of PD1 'Titans' and the later standard Guy double-deckers with Park Royal bodies. East Kent had a large number of Dennis single-deckers, buses and coaches, and a batch of Dennis 'Falcons' for one-man operated routes. Two worked on route 55 between Broadstairs and Kingsgate via the North Foreland lighthouse. During the height of the summer season, their 26-seat bodies were totally inadequate for the volume of traffic to the lighthouse and Joss Bay, and since the local police made every effort to prevent overloading an expedient was soon devised. The first 26 passengers to board paid their fares in the conventional OMO manner as they boarded; the remainder, with due regard to safety, crowded aboard and paid for their journey as they alighted. Thus, if they were obliged to alight before the destination because the forces of the law and

Gleaming in the summer sun, this 1937 Duple-bodied 'Tiger' TS7 of East Kent awaits an excursion load at Margate Harbour; originally used by the MT Company for London-Thanet express work the vehicle passed to East Kent in 1938.

[O. Alison

order were in sight, they only paid for what they had received; the system worked admirably and seemed to please all parties—except perhaps the poor bus which had not been designed for such strain. Occasionally a Leyland 'Cub' would substitute or duplicate for one of the 'Falcons' and we would be treated to a ride of Rolls Royce quality—smooth and quiet—there was certainly no impression that the vehicle was nearly 20 years old. One day in Margate, I discovered one of East Kent's 'odd un's'—a beautiful 'Tiger' TS7 with Duple coach body, one of five which had come to East Kent from the MT Company of Brockley. It was on an excursion stand at Margate Harbour with radiator shining in the sun and the rich East Kent red paintwork positively gleaming—the effect heightened by the tiny cream relief between the moulding. The driver of the vehicle was a vintage busman, proud to show off his steed to an admirer and happy to climb into his cab to be photographed for posterity. There may be flags sticking out of the top of the bus, Morris 8's peering from the back and a host of reflections in the mirror-like sides but this is one of my favourite pictures. In common with other operators of long routes, East Kent practised 'half way' changeovers and more than once in company with two loads of passengers, I have witnessed the solemn exchange of crews and vehicles apparently miles from anywhere. When on a later journey there would be no change one was apprehensive lest the driver would not know the way but these fears were dispelled as the journey progressed

Towards the end of 1954 came the news that the Hants & Sussex empire was crumbling and some bus riding was indicated. My interest lay in the section based on Horsham where, from a small garage (sic) at Roffey, a number of vehicles provided a modest pattern of routes. The basic services were by one-man-operated Bedfords with NCB bodies plying between Ewhurst and Crawley via Horsham. Two routes were involved but in most cases the buses worked from end to end, whilst the bus did a 'short' at Crawley by retracing its route from the town centre, and then turning into the area of the New Town known as Langley Green. If I recall aright, an hourly headway on the main route gave most of Langley Green a half hourly service into Crawley when the short operated, though when the bus was late this section was often curtailed. In Horsham itself, Hants & Sussex worked a double-decker jointly with London Transport route 405 to give an extremely frequent local operation between the town and Roffey Corner adjacent to the garage. This service was in excess of the demand and coupled with extremely low receipts on the long country route the company was in financial difficulties. The gradual growth of Crawley New Town was seen as possible salvation (not unlike Mr. Campbell at Basildon) but a determined rearguard action was of little avail and the best efforts of staff and technical advisers could not beat the inevitable. In the last weeks of operation fuel oil had to be purchased out of cash takings and the electric light and telephone had been dis-

Hants & Sussex Leyland PD1 with Northern Coachbuilders bodywork caught on the Roffey Corner-Horsham service long before the Receiver moved in.
[R. F. Mack

connected from the garage. The substitution of London Transport buses must have come as a great relief to the harassed manager but even the greater resources of London Transport were of little avail and after a chequered career part of replacement route 852 was handed over to Brown Motor Services and another section has been abandoned. Only at Langley Green is there now a thriving community, some of whose senior members recall the red single-deckers that weaved their way past the piles of bricks and ballast needed to build the New Town. Other Hants & Sussex operations were taken over by Southdown.

One organisation which had captured my imagination was the Mexborough & Swinton Traction Co. Ltd., whose trolleybuses could be seen fleetingly from the train between Cudworth and Sheffield. The earliest encounters on the ground were in Rotherham whence the company ran jointly with the Corporation on an interurban system, through the mining districts of Rawmarsh, Swinton and Mexborough to Conisborough. The fleet was then composed of Garrett 'Bull' BTH single-deckers with Garrett centre-entrance bodies, which had replaced the original Daimler and AEC vehicles of 1915 and 1922. The vehicles were of vintage outline sporting chrome fenders, tram car style push-in ventilators and bulb horns. Some of the breed had their trolley-bases arranged one on top of the other which gave a bizarre effect and also seemed to render them prone to dewirements. To offset this, a bamboo pole was carried on the roof where it could be reached with ease by anyone 10 feet tall or over. Finished in a maroon and cream livery, these vehicles were a severe contrast to the Rotherham fleet which was painted for "speed"; both operators were at that time confined to single-deckers because of low bridges. The departure barrier in Rotherham was clearly marked for routes 24 and 25, the Corporation identification for MEXBORO' and CONISBORO' and gave no indication that most of the vehicles to depart would be on routes A and B, the company identification. Joint operation was on a mileage basis, the corporation running as many miles under the company wires as the company did between the All Saints Square terminal and the borough boundary. The usual proportion of operation was one Corporation vehicle to 7 or 8 company vehicles. The Garretts, which gave a very solid ride at what seemed a monotonously low speed, were augmented by some refugees from Hastings and Notts & Derby during the wartime years when traffic was heavy, and further assistance came in 1943 when some

All Saints Square, Rotherham, with a 1929 Garrett 'Bull' 32-seat trolleybus of Mexborough & Swinton Traction receiving running repairs from the driver while passengers board; note the difference between the destination information on the bus and on the stop.

[W. J. Haynes

of the ubiquitous Sunbeam W appeared in single-deck form with Brush bodies. These vehicles were finished in plain grey but after the war the company went green and cream and invested in a smart new fleet of Sunbeam Brush 35-seaters. One visit to the Rawmarsh depot revealed that buses were also operated, and had been on and off since 1922. Their workings were of limited scope and were generally covered by secondhand vehicles or borrowings from nearby (and associated) Yorkshire Traction—in 1953 some petrol engined Bedford/Duple 29-seaters were to be seen and also a few ex-Devon General AEC 'Regals' with very smart Harrington bodies. Operationally, Mexborough was unusual as the timetables for the main services showed significant daily variations, frequencies being greater on Monday and Friday afternoons, whilst service K to Rawmarsh had 16 outward journeys from Rotherham before 9.8 a.m. followed by one at 10.53 p.m. and only eight morning journeys back to Rotherham. On Saturday afternoon and evening a 10-20 minute frequency operated, but Sunday saw only 3 early morning round trips and the late evening outward journey. Further evidence of originality was in the destination information. All the timetables and maps defined the termini at Conisborough as Conanby or Brook Square, but trolleybuses to Conanby from Manvers Main on Route C showed Conisborough High whilst the service to Brook Square from Rotherham (A) showed Conisborough Low. A journey over the main route was one of many contrasts through the steel works of Rotherham and the colliery villages, passing wheatfields between Rawmarsh and the Woodman at Swinton, winding through the narrow main street in Mexborough—where as early as 1931 an extension was opened to Adwick Road "because of the congestion caused by terminating trolleybuses"—on to the Old Toll Bar, where a small running shed was situated, and then skirting the Denaby Main and Cadeby collieries to reach Conisborough with its ancient castle. The all powerful bus replaced the last of the faithful trolleybuses in 1961 and has spun a web of routes around the original tentacles of trolleybus operation; nevertheless the memories remain.

Sticking to the trolleybus wires for a while reminds me of Hastings where the trackless reigned from 1928 to 1959. In 1939 there were eleven 'routes' (2-12) covered by a fleet of Guy vehicles. As already noticed, some of these found their way to Mexborough after Hastings received a batch of new Sunbeam double-deckers in 1940. With one exception the original fleet of eight open-top double-

En route for 'Conisborough High', a 1948 Sunbeam with Brush 32-seat body at the Old Toll Bar depot, Mexborough.
[O. Alison

deckers was scrapped, and the survivor languished behind the Bulverhythe depot for many years until it was refurbished and restored to greater glory as an illuminated bus. It now serves as a petrol bus and must be unique in view of its age and history. The post-war years brought a further 25 Sunbeams to make a total of 45 plus the Guy double-decker. Two or three of the Guy single-deckers remained for many years but quietly disappeared from the Silverhill depot before the system closed in 1959. In post-war years the original eleven routes had come down to four, though only a small amount of 'wireage' at Blacklands was not in regular use. The Circular Route 2 was unchanged; route 6 retained its Hollington terminal but was projected from the original Langham Hotel terminal to serve Ore and St. Helens, replacing sections of 7 and 12. Route 8 from Alexandra Park was projected from West Marina (Bo Peep) to Bexhill and Cooden, previously served by 5, 7 and 12; and route 11 covered its original ground between Hollington and Ore with alternate projections to St. Helens. In pre-war days the Circular route enjoyed a 15 minute headway but by 1956 this had dropped to about 40 minutes; on remaining routes there was a widening of headways with the pruning, but it must be remembered that the pre-war service was basically single-deck whereas post-war operations were entirely double-deck. Pre-war routes 6, 8 and 10 appear to have been provided with a view to catering for the holidaymaker as they did not start up until about 10-0 a.m. and provided the same facilities daily.

In post-war years this was a very snappy system, with well maintained vehicles and clockwork regularity. Only the drab green suggested that there may be a connection with Maidstone & District next door (they were in fact connected as early as 1935) and for many years quite separate organisations were maintained in Hastings. The trolleybus superintendent was responsible for a large and well equipped works, which dealt with nearly all overhaul items and in addition turned out considerable quantities of metal castings, mirrors, brake pedals and the like to be used on the vehicles of the parent company. One other feature of the Hastings trolleybuses is unique; in Western Road, Bexhill, there was a feeder pillar, clearly labelled "Swinton Corporation Tramways" complete with coat of arms, and another at Blacklands had seen its early years at either Mansfield or Musselburgh. Very few bore any identification to connect them directly with Hastings. For many years insert Setright tickets were used (another connection with M & D?) but latterly the speed model took over. Also in Maidstone fashion, the route numbers were circled 8 etc. and a further number was carried on a metal plate fixed in front of the steering wheel, but the object of this was never clear. In many cases cast fleet numbers were used, no doubt the products of the metal shop at Silverhill depot. The trolleybus was, of course, well equipped to cope with the steep hills that abound in Hastings and their silence will never be equalled by the Atlanteans that took over on many of the routes from June 1959. Whether for acclimatisation or lack of vehicles I know not, but in the last weeks of trolleybuses it was not unusual to find that an Atlantean had been substituted for the trolleybus but this was no doubt to be expected since in their last three years the trolleys were labelled "Maidstone and District" instead of "Hastings Tramways". The ability of M & D to turn out a 35-seater coach on a working which really calls for a 78-seater, when other double-deckers are apparently available, never ceases to amaze me. Perhaps it is part of a cunning scheme to confuse the casual passenger as the substitute vehicles are seldom clearly identified, they just materialise.

A good base for the county of the Red Rose is Bolton where Ribble rubs shoulders with Lancashire United and buses from Bury and Salford may be seen alongside those of the Bolton Corporation. From Salford's Victoria bus station in Manchester (or is it Manchester's Victoria bus station in Salford)? route 8 of Bolton, Salford and LUT provides a convenient link to Bolton serving places like Smithy Brow, Irlams O'th Height, Billy Lane, Unity Brook, Man and Scythe, and Moses Gate, and in an area where rumour has it that the scene is uninteresting some magnificent views are to be seen. On one such trip the Bolton Wakes Week was ending and it was amusing to see families raking out purses and pockets to find the bus fare home after their spree. On another occasion the whole area was in the grip of a drought and normal methods of washing vehicles were out of commission. The Bolton fleet was immaculate, however, as the result of some real hard work with damp cloths; this was before the present livery was adopted and the amount of cream paint was limited, but those buses really gleamed. Red in one or other its interpretations was the standard colour for four of the locals and the respective greens of Bury and Salford provided a welcome contrast. An even greater contrast was available if one took the train to Lytham where the dark blue of the multi-Leyland fleet seemed to be in keeping with the atmosphere of the twin towns. My favourites here were the full-fronted double-deckers which well nigh purred along between the dunes as they made their way towards Squires Gate and the glitter of Blackpool. The all-Leyland product, with the GEARLESS BUS legend on the radiator, was, of course, very close to its birthplace and their single-deck 'Lion' contemparies completed the picture. The cessation of bodybuilding by Leylands must have caused some managerial mutterings in Lytham. The bus operations of that well known northern neighbour of Lytham St. Annes seem to come a poor second to the famous tramcars, but there is no disputing that the bus now has the back streets to itself. In Blackpool the Leyland is all powerful and in keeping with the modern outline of the tramcars the buses have sported full fronts for many years—both double and single-deckers—and for an even longer period centre entrances held sway. In direct contrast to the trams, the bus fleet had very generous destination blinds; the proportions have now been adjusted to give more room to the destination and less to the number and the entrance has slipped back to the rear. In the summer, seats are taken out of some vehicles to provide more luggage space—perhaps one day they will leave the seats alone and put a trailer on for luggage. No doubt with an eye to their visitors, Blackpool have always provided an excellent map of services combined with a full street plan; when read with the timetable the wealth of information given is almost overwhelming.

The return to Bolton on this occasion was by the quarter-hourly X60 service of Ribble, North Western and LUT which allegedly takes about 100 minutes. The North Western

Above. This Lytham PD2 heads homeward under a festoon of Blackpool tramway overhead.

Below. In Blackpool bus station—a 1950 Guy 'Arab' with Northern Counties body incorporating coach-type seats for limited stop operation.
[R. F. Mack

Bristol which conveyed your scribe was totally unsuitable for an express service except in its speed potential, thus the agony of cramped massage was reduced by 10-15 minutes and I was stretching my limbs in my hotel room when we should have been drawing in to Bolton's Moor Lane bus station.

A different approach to Bolton is from the north and I recall one trip which started after a train journey from Edinburgh to Hellifield thence to Blackburn. From Blackburn, where the Guy-East Lancashire product was gradually ousting the traditional Leyland, the journey was over the old tram route into Accrington where again the Guy appeared to prevail. The journey was made on a Saturday afternoon and a study of passing vehicles gave the impression that all Accrington was intent on going to Blackburn whilst the loadings of our steed and the hopeful queues showed that most of Blackburn and at least half of the western populace of Accrington had their sights set on Accrington Market. A pause in the centre of Accrington helped to dispel the theory but set one wondering how it was that Accrington came upon its distinctive livery of dark blue, black and red. South from Accrington and Rawtenstall, Leyland was the conveyance, and thoughts turned to the manager of the undertaking who also runs the Haslingden and Ramsbottom fleets but sympathy waned a little when simple addition revealed the combined fleet strength to be around 80, a figure exceeded in a single garage of many other operators. After sampling a Ramsbottom single-decker (Leyland, of course) into its home ground, a Ribble flier speedily completed the journey to Bolton.

Whether because of modesty or some other more compelling reason Bolton issued neither timetable nor map, and one had to rely on a miniscule publication which showed departures only, so that the compilation of connections was, to say the least, difficult. Combined with this, the LUT 'arranged' its routes in "numerical" order in its own book—that is to say one number after another—29, 30, 18, 28, 57, 58, 54, 60, 61 and 79 for example with 1 bringing up the rear! Headaches were often the reward for timetable searching and there was a tendency to rely on the well tried routes. One such was the South Lancashire trolleybus service to Leigh which drove south west from Bolton. The first three miles to Four Lane Ends, Hulton, were in the Borough of Bolton and for the first three years of the trollybus route, Bolton maintained a parallel tram service. In 1936 this was also abandoned and Bolton became the technical owners of four trolleybuses which were in SLT livery. A complicated system ensured that Bolton collected its share of the receipts, each conductor having two ticket machines which were religiously transferred at Four Lane Ends whereafter the conductor made a second collection round the bus. The 'Bolton' trolleys were Leyland-Roe 64-seaters and latterly they were to be found on other parts of the system. After crossing the 'trunk' route in Atherton, the Leigh route passed between the various headquarters garage and power station buildings at Howe Bridge before reaching the centre of Leigh where the terminal was adjacent to an excellent fish and chip shop. The Leigh route always seemed to command the latest vehicles, though not necessarily the best in condition. Since conversion to buses in 1958 the Leigh route has been jointly operated with Bolton and the re-booking ritual eliminated.

The obvious sequel to the Leigh trip was to venture on the Farnworth to Atherton trolleybus section. This 72-minute ride was undertaken one Sunday morning (very damp) when it was easy to see why a 30 minute headway sufficed until midday. Because of the circuitous route, through passengers were a rarity and were regarded with suspicion by the conductor—at Little Hulton the route was only a little over a mile from itself at Mosley Common but took 42 minutes to get there! At Swinton a four minute stand was allowed and then progress was resumed to Atherton where, save for the relief provided by Worsley Court House, little was offered in the way of scenery or architecture. The conveyance was one of the original 1930 Guy Roe vehicles which had been rebuilt by Bonds to give a more modern outline than the 1930 "piano" front and it says a great deal for the makers and maintainers that many of these vehicles lasted throughout the 28 years of operation.

The conclusion of the journey was a wild dash to Manchester on a Guy-Northern Counties double-decker of LUT giving no grounds for comparison with the earlier stately progress of the trolleybus. The walk to Piccadilly confirmed that nothing ever changes—Manchester had a new 'sameness'—different from the Crossley image of 1945 but just as uniform in its new way and visually much less attractive. Whatever the economies effected by monochrome finish, to my mind they can never be justified when the unadorned vehicle looks drab to the point of being almost repulsive.

The final recollections concern Eastbourne where the idea of municipal bus services first saw the light of day. Today, the blue and primrose Corporation buses run side by side with the pleasant apple green and cream of Southdown, though the peace is occasionally disturbed. Compiling notes of journeys made fleet lists and so on, draws attention to links with other areas and strangely enough Eastbourne has a connection with Lancashire United as fifteen buses from Eastbourne worked for LUT on loan during the war. With around 50 buses, the Corporation provides a network of services which divide into two groups, one running parallel to the coast between the foot of Beachy Head and Langney, the other passing the station and serving the Old Town and the rapidly developing area at Hampden Park. A seafront service is operated starting from the Railway Station, through Meads to the foot of Beachy Head, thence along the front to

sections, and on doubtful days alternate open and normal buses were used; substitution of the open buses took an hour if bad weather really set in and so a batch of convertible double-deckers with full-drop windows on the top deck was put into use, also in the white livery.

Another claim to fame by Eastbourne was that until 1946 they had no oil engined buses and the petrol vehicles survived well into the 'fifties giving wonderfully silent rides. Because of the high proportion of seasonal work longevity of the fleet is the rule and the long fallow period of winter facilitates maintenance. Distribution between Leyland and

The first of the South Lancashire trolleybuses built in 1930 with a Roe 53-seat lowbridge body on a Guy BTX chassis depicted—before rebuilding with a straight front—at the Farnworth terminus.

(R. F. Mack)

Princes Park and a terminal near the Churchdale Road garage where the fleet is housed. This service was operated by white painted decapitated double-deckers and started in 1947 with some 'Titan' TD2's dating from 1932. Their Leyland bodies were converted by the Corporation and, free from external advertising, they provided an attractive ride for holidaymakers and enthusiasts alike. Each bus was named—White Queen, White Princess, White Lady and White Rabbit forming the original fleet. As the years passed by, the TD2s gave way to TD4s, then to a batch of 1938 'Regents' with Leyland engines and Northern Counties bodies, until the present fleet of 5 PD1s with East Lancashire bodies was converted from 1961. The vagaries of the English climate gave rise to allocation problems on this route, which provided a normal local service over some

AEC was almost equal in early post-war years though a batch of eight Crossleys slipped in around 1949, but all recent purchases have been of the AEC marque with East Lancashire bodies. The fleet includes two single-deckers, one 'Regal' and one 'Lion' which are seldom seen in public service but do a certain amount of school and contract work. A shortcoming so far as I am concerned is the habit of plastering the whole of the upper deck with painted advertisements which make it difficult to determine which words are related to beer and which to the destination. The destination display also leaves something to be desired but this is offset by a very well produced folder map and timetable.

As well as operating the longer stage services around Eastbourne, Southdown provide the service (97) to the top of Beachy

Above. A 1936 all-Leyland TD4C leaving Eastbourne Pier; originally No. 94 in the Corporation fleet, it is now called 'White Heather'.

Below. Having coasted down from Beachy Head, Southdown 446—a 1945 Guy 'Arab' with Park Royal body—pauses before proceeding to the Royal Parade terminus of route 97.
[*O. Alison*

Head. This is very popular and the original fleet of six-wheel Leyland single-deckers gave way to open-top Guys converted from war-time utility buses. The characteristic roar of those stalwarts has now been overtaken by the equally resonant tones of some 69-seat convertible 'Titan' PD3s. The success of the 97 route led to its extension as 197 to Devils Dyke giving a circular ride through the splendid Sussex countryside. The Leyland product has long been top dog with Southdown though there was an attack of Guys during and after the war together with a handful of Bedfords and Dennis. Unhappily the roar of the 'Titan' TD is no more, neither are the silent 'Cubs' and 'Cheetahs', but their successors bear the Southdown name just as proudly on their gleaming flanks.

The only "rift in the lute" at Eastbourne is that from time to time the Corporation cast envious eyes at the traffic going to the top of Beachy Head, an area owned by the Corporation as well as being within the borough. So far their efforts to obtain a licence for a service have met with failure as Southdown ensure that their own operation is capable of meeting the demand and the Corporation are unable to prove that the existing facility fails to satisfy the public, or that they have a greater right to the traffic. It is all on a very gentlemanly basis, of course, as these matters should be. As all things are possible we may yet see white double-deckers labouring up the slope to discharge their loads alongside the enormous PD3.

The compilation of these meanderings has brought back many pleasant memories of incidents and friendly busmen and I hope that it will provide pleasure to my readers, if only to stimulate them to travel more by bus, to make their notes and comparisons, to harbour likes and dislikes and then to savour their recollections as I have done.

Good bus riding!

Lancashire-Scottish Express Services

by A. J. Douglas

SURPRISINGLY, many people are completely unaware of the existence of the network of express coach services covering this country; yet, in these days, when the bus industry generally is suffering from the effects of declining passenger travel, these express services are actually attracting more passengers every year. The services between Lancashire and Scotland make up one of the longest-established and best-patronised groups of such services.

As early as 1919, W. Alexander & Sons Ltd, ran their first tour from Falkirk to Blackpool using a Leyland charabanc. This was a six-day affair taking two days for the journey in each direction, with overnight stops at Carlisle, and allowing the holiday-makers two days' stay in the Lancashire resort. The outward journey was made by way of the direct route via Beattock and Shap, and the return journey followed the more scenic route through the Lake District, the Tweed Valley and Edinburgh. The tour was a great success and ran for a number of years, mainly at the time of the Falkirk and Glasgow Fair holidays.

It was not until 1927 that the first regular express service started when, in April of that year, Anglo-Scottish Motorways commenced operating between Glasgow and Liverpool using four W & G coaches and two Crossleys. The service was timed to connect with Scottish Clan Motorways' Glasgow-Aberdeen service which started at the same time, and a connection was also made at Liverpool with a service to London. Of course, several difficulties were encountered in those early days, not least being the totally unrealistic 12 m.p.h. speed limit imposed on all buses and coaches weighing more than two tons. In June 1927, at Ormskirk, Anglo-Scottish Motorways were fined £3 on each of two charges, and the two drivers 10s. each, for exceeding the limit, the police stating that the Glasgow-bound coaches were travelling at 35 m.p.h.!

The potentialities of express service operation across the border were soon realised by other operators, and other services soon started up. Lowland Motorways, of Glasgow, introduced a Glasgow-Manchester service in June 1928, which connected with the Man-

Above. Typical of the modern vehicles on the express services between Scotland and Lancashire in the summer of 1965 is Ribble 789 at Canderside Toll on route X30 bound for Manchester.

[all photos by the author

chester-London service operated by Tognarelli, of Bolton, and later by Fingland of Manchester; a much more ambitious connection was also made with a Yelloway service to Torquay. A Manchester firm, Weaver & Greatorex, renamed W & G Coachways (Manchester) Ltd. in 1933, also operated between Manchester and Glasgow, In this case, day and night services were provided all the year round. At one time Mr. Moller, of Lowland, approached W & G with a view to co-ordinating the two services but negotiations proved fruitless, and Lowland thereafter concentrated on providing a service during the summer months only.

Glasgow-Liverpool services were provided by two operators from the Lancashire city—Imperial Motor Services and McShane's Motors Ltd. The latter was sold to Red & White Services Ltd., of Chepstow, in 1933. A Hawick operator, Adam Graham, ran a service from his home town to Blackpool.

Naturally the major operators were not left out of these developments. In the spring of 1930, services to Liverpool, Manchester and Blackpool were commenced by the Scottish General Transport Co. Ltd. and Ribble Motor Services Ltd., operating in partnership. The starting point was Paisley, where Scottish Transport had a garage at that time, and the coaches passed through Glasgow, where they picked up passengers at George Square. The Paisley-Glasgow section was withdrawn in 1935, probably due to the closure of the Paisley garage. Similar services were started from Edinburgh at the same time by Ribble and the Scottish Motor Traction Co. Ltd. The Liverpool services connected with a service to London operated initially by the Merseyside Touring Co. Ltd. and then by Crosville. The fare at this time was 20s. single or 31s. return between Glasgow and Liverpool or Manchester, and 19s. 9d. or 32s. from Edinburgh.

The SMT subsidiary, Midland Bus Services Ltd., whose Glasgow-London service is perhaps better known, also ran a competing service between Glasgow and Blackpool. Scottish Transport at this time was a subsidiary of the British Electric Traction Co; however, in 1932, it was acquired by SMT and was renamed Western SMT. Shortly afterwards the Midland company was absorbed into the new company and the Blackpool service was merged with the former Transport service.

The Road Traffic Act of 1930 (which did not take effect until 1931) imposed an almost insuperable stumbling block on the launching of any further competitive services and a period of consolidation and mergers followed. Lancashire United applied for services from Manchester to Glasgow and Edinburgh in 1931 but the application was withdrawn. Then, in 1934, the W & G Coachways and Lowland Motorways services were purchased by Ribble, but Western SMT did not participate in the Manchester-Glasgow night service thus acquired. The next moves saw the Liverpool-Glasgow services of Imperial and Red & White passing to Ribble in 1934 and 1935 respectively; and also in 1935 SMT acquired Graham's Hawick-Blackpool route, which was incorporated in the Edinburgh-Blackpool service. So that by the outbreak of the second World War, Ribble and SMT enjoyed a monopoly of the Lancashire-Scottish services.

The services were suspended from 1940 until the end of 1945, when a start was made on gradually building them up once again. In 1952 an Edinburgh Liverpool night service was started. This operates during the summer only but, by connecting with the Glasgow-Manchester night service, it provides overnight facilities between the two Scottish cities and both the Lancashire cities. Another important development in 1952 was the formation of the 'Westlinks' network. This provides connecting services between Scotland and the South-West of England and South Wales by linking the Scotland-Liver-

Ribble's night coaches do not always lie idle in Scotland during the day—No. 936 at Kilmarnock after bringing a load of Rangers F.C. supporters in from Glasgow.

pool services with the Associated Motorways network from Cheltenham by way of Ribble's Liverpool-Cheltenham service.

In February 1951, after a long and bitter battle in the licensing courts, Northern Roadways, a Glasgow independent operator, was granted licences for express services from Glasgow and Edinburgh to London. Certain restrictions were placed on the licences and the fares were considerably higher than those charged on the SMT and Western SMT London services (40s. single and 75s. return, as against the Bus Group's 30s. single and 60s. return at that time) but a completely new standard of comfort was set. The Burlingham 'Seagull'-bodied Leyland 'Royal Tigers' and AEC 'Regal' Mk. IVs were fitted with toilets, plugs for electric razors and hot water tanks; hostesses were carried to attend to the passengers' needs and serve refreshments en route. Despite the fares differential this competition had a decidedly beneficial effect, and SMT and Western hastened to introduce new AEC 'Regal' IVs with specially designed Alexander bodies, also fitted with toilets, for their London services. Unfortunately, the benefits of these exceptionally comfortable vehicles did not permeate through to the Lancashire services.

Naturally, the existing operators (and British Railways) appealed against the grant of these licences, but in the meantime, Northern Roadways applied for, and were granted, two further licences in the autumn of 1951. One of these was for a service from Glasgow to Bournemouth; but the other, from Glasgow to Birmingham, is of more direct interest to us in this article. It was bound to lead to abstraction of traffic from the Lancashire services, many of whose passengers caught connections to the Midlands. Appeals were made against these too. At last, in March 1952, the Minister of Transport announced the appeal decision—all Northern Roadways' express licences were to be revoked. There was a storm of protest; the story made the headlines in the daily papers; the advantages of competition were already evident on the London services; and, of course, there was the typically British support for the underdog! As a considerable number of passengers had already booked with Northern Roadways, the Minister allowed the services to run until September 30. The objectors protested strongly over this unprecedented concession, but it was allowed to stand.

Northern Roadways, of course, did not take this lying down. They promptly re-applied for the licences and, after two stormy hearings in September and November, a further grant was made, although the West Midland Traffic Commissioners, when granting the backing licence, prohibited the carriage of passengers whose journeys originated in Birmingham on the Birmingham-Glasgow service. Once again the objectors appealed, but this time they were unsuccessful and Northern Roadways appeared to have established a secure foothold in the

Note the route number — 30X — on Western SMT's ML1887 which was about to leave Glasgow for Manchester in January 1966.

express service market. It is sad to have to record that in 1956 Northern Roadways gave up the struggle and sold their express services to Scottish Omnibuses Ltd., the licence for the Birmingham service being granted to the latter in November.

Meanwhile, Ribble had not been inactive. Stung to action, no doubt, by the grant of the Birmingham licence to Northern Roadways, they had applied in June 1951, jointly with Midland 'Red' and North Western, for a new licence extending the Glasgow-Manchester night service to Birmingham. This application was withdrawn, but a fresh application was made by the same three operators in May 1952 to extend the service, this time to Coventry, and this was eventually granted 18 months later, in November 1953.

When Scottish Omnibuses were granted the former Northern Roadways Glasgow-Birmingham licence, the same three operators promptly applied for a similar service. Obviously, some agreement was reached with the Scottish operator, for in February 1957, the three companies were granted licences to operate the service jointly with SOL and the following month Western were licensed to run jointly with Ribble, North Western and Midland 'Red' on the Coventry and Manchester night services. Now, presumably, all concerned relaced a little once more. However, in the winter of 1962/63 the cat was put among the pigeons once again.

A Coventry operator, George McGhie, applied for an excursions and tours licence to operate from Coventry to Glasgow one weekend per month and at Christmas, Easter and Whitsun weekends and at the local industrial fortnight. This seemed a small request but it would have given McGhie a foothold in a potentially very lucrative market. The major operators, having made a considerable investment in developing their Coventry-Glasgow service— an investment, indeed, which had not yet been recovered—objected strongly.

McGhie produced an impressive array of witnesses at the traffic court hearing. It was pointed out that, for the large community of Scots working in Coventry, the present arrangements were most unsatisfactory; anyone wishing to use the service for a weekend visit to Glasgow had to leave work early to catch the coach on Friday afternoon and if, he used the Sunday night services for his return journey, he did not reach Coventry until midday on Monday. By operating direct, McGhie proposed to offer much more suitable departure and arrival times. The traffic Commissioners were sympathetic, but they felt that a service of this nature should be covered by an express, rather than an excursions and tours licence, and accordingly advised McGhie to make a new application.

In May 1963 the revised application was heard. This time the major operators also submitted an application for a direct service from Coventry to Glasgow, stopping to set down and pick up only at Birmingham, Wolverhampton and Hamilton, and utilising

Ribble 759 in Bolton on the X50 service from Morecombe to Manchester, which parallels the Scottish services into the Lancashire city.

the new section of the M6 motorway. This was granted to operate daily from Whitsun to September and at weekends and holidays during the rest of the year. McGhie, on the other hand, received only a very limited licence to operate to Glasgow on the first Friday of the industrial holiday, returning on the second or third Sunday. Even at this, a restriction of three vehicles was imposed and higher fares were enforced. Thus, having shown the initiative and explored the market, McGhie appears to have had a rather raw deal. However, it must be borne in mind that Ribble and partners had laid out a considerable amount of money on developing their service via Manchester, which had frequently operated at a loss, and they had intended to improve the service once the M6 was fully opened.

In January 1966 another operator was granted a licence for a night express service between Birmingham and Scotland. This was Stockland Garage Ltd., of Birmingham, and the service operates not to Glasgow but to the Ayrshire coast resort of Largs, with a setting down and picking up point at Butlin's holiday camp in Ayr. The service was authorised to operate with a vehicle allowance of one from April to September during 1966 only. Witnesses for Stockland said they wanted to travel to Largs without having to change, and they also wanted to book their holiday hotel accommodation with their coach seat. Witnesses supporting the application for the stop at Ayr said they wanted to spend holidays at Butlin's, but had not done so due to the necessity for changing to the stage service to Ayr at Glasgow, and Stockland's representative pointed out that Ayr was the only Butlin's camp not to have a direct service from Birmingham. The grant was made subject to the condition that Stockland kept records of the number of passengers using the service and the number of those going to Largs who also required hotel accommodation.

One other service which should perhaps be mentioned here is Barton's route 58 between Corby and Glasgow. A night service is operated leaving on Fridays and a day service on Sundays, both during the summer season only. The service is intended mainly to help the large Scottish community in the Midlands steel town, but the northbound Friday service and the southbound Sunday service stop in Nottingham and Leicester. Although this does not compete directly with the Lancashire service, it is bound to attract some passengers who would otherwise use the Nottingham-Manchester express operated jointly by Trent and North Western, which connects with the Scottish services.

The following list should be read in conjunction with the accompanying map of services at present operated.

(Operated by Ribble and Western SMT)

Ribble Route No.	Western Route No.	
X_1	97D	Glasgow-Liverpool (Day service only)
X_2	97C	Glasgow-Blackpool (Day service only)
X_{20}	97B	Glasgow-Manchester (Night service)
X_{30}	97A	Glasgow-Manchester (Day service)

(Operated by Ribble and Scottish Omnibuses

Ribble Route No.	SOL Route No.	
X_{10}	500	Edinburgh-Manchester (Day service only)
X_{11}	501	Edinburgh-Liverpool (Day service)
X_{12}	503	Edinburgh-Blackpool (Day service only)
X_{51}	502	Edinburgh-Liverpool (Night service)

(Operated by Ribble, Western SMT, North Western and Midland Red)
  Glasgow-Coventry via Manchester
  Glasgow-Coventry direct

Of the above, only routes X_{11}, X_{20}, X_{30} and the Glasgow-Coventry service via Manchester operate all year round. The remainder are summer services, with the direct service to Coventry also operating on certain days during the winter. However, this does not give the full picture. Connections with the Glasgow-Manchester night service provide links to Liverpool and Blackpool; similarly, connections for Manchester and Blackpool meet the Edinburgh-Liverpool night service, but in practice, when traffic warrants, through coaches are run. The X_2 operates via Kilmarnock and Dumfries, whereas X_1, X_{20} and X_{30} go by way of Hamilton and Lockerbie. Again, where traffic warrants, X_1 and X_{30} duplicates can be routed via Kilmarnock and Dumfries. Presumably these are technically running on

THE LANCASHIRE-SCOTTISH EXPRESS SERVICES 51

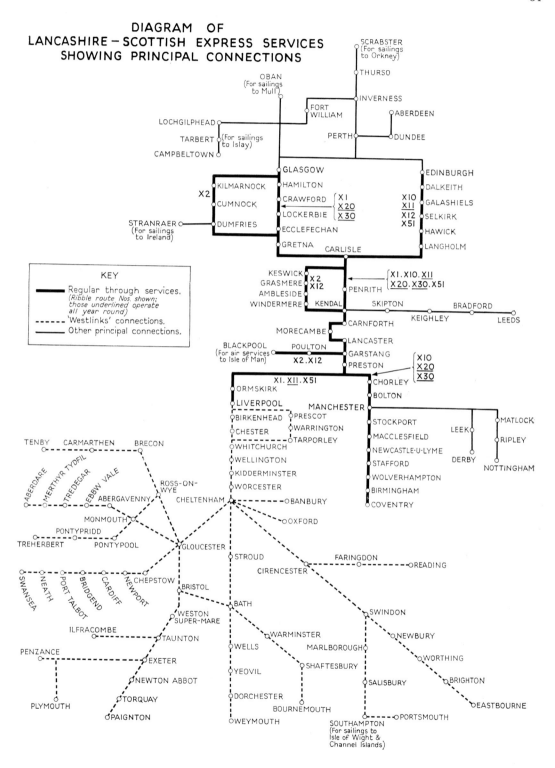

On hire for the Blackpool service—Standerwick 123 at Penrith in August 1958.

X_2 until Carlisle and changing to X_1 or X_{30} there. The X_2 and X_{12} services take a roundabout route via Keswick and Ambleside from Carlisle to Kendal, instead of the more direct route through Penrith followed by the other services. Duplicates, however, frequently take the Penrith route.

As a result of the above arrangements, a much more comprehensive service is provided than is apparent from the published timetables. The observer watching the departure from Glasgow at 22.20 hrs. on a busy evening will see coaches bearing a multitude of destinations—Preston, Liverpool, Blackpool, Morecambe and Birmingham, as well as Manchester and Coventry. All Ribble vehicles, including the Coventry coaches, carry the route No. X_{20}. Western coaches, if they carry a number at all, usually show '20X', the nearest they can get to the Ribble route number. Rarely does a Western coach show its own route number '97B'.

On long journeys of this sort, refreshment stops are essential, both for the sake of the passengers and to give the driver the breaks required by law. The principal stop, of some 40-50 minutes, is made, naturally enough, about half-way. Apart from the Blackpool services, which stop at Carlisle on the southbound journey and Keswick on the northbound, and the direct Coventry service, which stops at the M6 motorway cafe at Charnock Richard, this stop is made at Penrith. At busy times this can cause some trouble since the restaurants and cafes in the town just cannot cope with the large numbers involved. The situation is particularly bad at night when there is only one small cafe open. As there are quite often some 40 or more coaches parked outside at one time, the ensuing confusion is not difficult to imagine. The sight of all this activity in the square which serves as a bus station and car park during the day, is nonetheless an interesting one to the enthusiast. On one recent journey the author made from Manchester to Glasgow an enterprising driver carried on for a few miles beyond Penrith and stopped at a much better equipped transport cafe. A few other coaches were parked outside, so presumably, this is a fairly regular occurence.

The operation of these services by the different firms varies a little. Western operate the regular coaches from their Newton Mearns garage; relief vehicles and crews are drawn from any of their other depots. Scottish Omnibuses provide the service vehicles from their New Street depot in Edinburgh, but duplicates are also supplied by Dalkeith, Musselburgh, Broxburn, Bathgate and Linlithgow. All the crews involved do stage carriage work regularly and a turn on the Lancashire services is something of a 'plum' for them. Ribble, on the other hand, employ a number of drivers specifically for these duties, including a number based in Scotland. Ribble crews on the service vehicles change over at Penrith to allow them to return to their base. Duplicate crews, of course, frequently travel all the way, returning the following day or night. The Glasgow-Coventry via Manchester service operates virtually as two services using the same coach. The Glasgow-Manchester section is run as already described, but at Manchester a Ribble or Midland 'Red' crew takes over for the journey to Coventry.

An unusual feature of the Lancashire services is the carriage of conductors. This is a long established practice but the conductor, on the night service particularly, seems to have very little to do. On the day service more intermediate passengers are carried, especially on the day services from Edinburgh which operate as stage carriage services between Hawick and Carlisle (between Edinburgh and Carlisle during the winter), but the driver could still cope quite comfortably. It seems to be normal practice for the driver to help the conductor with the loading and unloading of passengers' luggage, so the only additional work involved, if conductors were dispensed with, would be the issuing of tickets. However, the method of doing this seems unduly complicated. Three books of tickets are carried. One type is issued as a return ticket when the voucher issued by the advance booking office for a return journey is handed over. A similar single ticket is given in exchange for a voucher for a single journey. When a return voucher is surrendered an exchange ticket is given. All these tickets have to be written out by hand—and, incidentally, carry Ribble's name, regardless of the ownership of the coach. Many passengers do not book in advance but pay the conductor. They receive similar tickets. Most other coach services appear to operate quite satisfactorily with a less complicated system whereby the lower portion of the advance booking voucher is torn off and returned to the passenger. When a passenger pays cash, the driver makes out a form similar to that used for advance booking.

Changes are being made, however. For many years, coaches on the night services from Edinburgh at least have carried only one conductor for every three vehicles. As from November 1, 1965, SOL introduced a new type of pre-booking voucher which has a detachable portion for return to the passenger, instead of the tickets already mentioned. With the introduction of these, only the service coach carries a conductor. Passengers paying cash on the service coach receive

SOL's Bristol A22 between Thurso and Wick en route from Scrabster to Edinburgh—technically the Highland Scrabster to Inverness service—to connect with the service to Liverpool.

Setright tickets. Western also hope to dispense with conductors whenever possible in the near future.

There are numerous connecting services but only the most important of these can be shown on the, map Some are worthy of mention here. The 'Westlinks' organisation has already been mentioned. The Ribble X24 service from Liverpool to Cheltenham links the Glasgow-Liverpool service with the Associated Motorways network from Cheltenham. Ribble also operates another service, X25, through Cheltenham to Bristol. The services from Glasgow and Edinburgh to Inverness shown on the map connect with Highland's Inverness to Scrabster service, but in practice, the Edinburgh-Inverness night service coach frequently works through to Scrabster. The West Yorkshire connection from Kendal to Leeds and Bradford meets the night service as well as the day service. At busy times, coaches work through from Leeds to Glasgow, operating on hire to Ribble from Kendal.

There is no real competition from other coach operators—McGhie's and Stockland's services, and Barton's Corby-Glasgow service can have little real effect on operations of this magnitude. However, there is still a fierce struggle for passengers with British Rail and the airlines. The coach's main advantage is its lower cost. The following examples of fares from Glasgow give some idea of the advantage this represented in March 1966.

	Coach Fare	Rail Fare (2nd Class)	Air Fare
	Single Return	Single Return	Single
	s. d. s. d.	s. d. s. d.	s.
Manchester	33 9 53 9	59	98s. (Sat. Sun. 75s.)
Liverpool	33 9 53 9	58	95s.
Blackpool	30 3 49 9	56	Service withdrawn
Birmingham	45 9* 74 9*	73	111s. (Sat. Sun. 90s.)

(*Direct route. Fares are 46s. 9d. and 76s. 9d. respectively via Manchester)

Ordinary return fares by rail or air are twice the single fare in each case. The first thing to strike one is that in every case, except Birmingham, the return fare by coach is less than the single fare by rail. However, influenced no doubt by the competition, British Rail offer cheap weekend excursion fares of 63s. to Liverpool and Manchester and 79s. to Birmingham, but these are not always available during the peak summer months. The air services to Manchester and Birmingham are operated by BEA, and the Liverpool service by British Eagle. Starways used to offer a service to Liverpool via Blackpool at a much cheaper fare but, since this airline was acquired by British Eagle, the Blackpool facility has been withdrawn and the Liverpool fare increased to bring it into line with BEA's fares.

It is of interest to consider what extra benefits British Rail and the airlines offer for their higher fares. First of all, both are quicker; the coach to Manchester, Liverpool and Blackpool takes something like 10 hours compared with the fastest BR express times of some 5 hours, and the flying time of one hour (not to Blackpool of course). The few hours saved do not justify the extra cost to many people—and many prefer the coach journey in any case. To the flying time of one hour must be added the half-hour or so to allow for checking in plus the time spent in travelling to and from the airports; the actual time involved then becomes much more comparable with the rail time. When the overnight journey is considered, the time saved by travelling by train becomes less as, to avoid unduly late departures or over-early arrivals, (as it is the trains arrive around 06·30 hours) the time is increased to some $6\frac{1}{2}$ to 7 hours. Furthermore the Manchester coach does the night journey in only 9 hours. As far as comfort is concerned, the train appears to offer little advantage over the coach unless a sleeping compartment is booked at a considerable extra cost.

On journeys to Birmingham and Coventry until the introduction of the direct service, the coach was at rather more of a disadvantage, taking over 13 hours to Birmingham and more than 14 to Coventry. The direct service, when it is operating, cuts these times to under 10 hours and $10\frac{1}{2}$ hours respectively, which compares with the rail time to Birmingham of 6 hours (7 by night train) and a flying time of one hour ten minutes.

Many people, of course, travel by their own car. The effect of this is much more difficult to assess. Petrol costs for a return journey to Manchester or Liverpool would be at least 50s. This, of course, is independent of the number of passengers in the car. Against the convenience of having the car available at one's destination must be set the strain of driving all the way; and the petrol costs, as many car owners seem to forget when costing such journeys, account for only a fraction of the real running costs.

The coach operators meet all this competition very successfully. However, it is possible that they could do even better. First of all, much could be done to improve their image. Many people seem to consider it a bit 'infra dig' to travel by coach. Yet, why should this be so? Properly organised advertising could surely do much to improve this situation. Advertising could also bring the existence of the services to the attention of the vast number of people who know nothing whatever about them. Certainly some excellent publicity material is produced at present but this is mainly in leaflet form and so is seen largely only by those who are already interested. The expenditure of some money on TV advertising, for instance, would probably be amply repaid.

The vehicles employed on the Lancashire services have always provided plenty of variety. Unlike the Glasgow and Edinburgh to London services, these Lancashire routes are not considered long enough to warrant the maintenance of a special fleet of coaches. Just about every type of coach owned by the various operators concerned is used, but they are normally fitted with 'dimmable' interior lights to give passengers a chance to sleep during night journeys.

Ribble, SMT and Scottish Transport all started the services using Leyland 'Tigers'. The Ribble machines had coach bodies built

THE LANCASHIRE-SCOTTISH EXPRESS SERVICES 55

Coaches on the Lancashire services are truly dual-purpose; Western SMT's KL527 was used with equal frequency both for express duties and on local services in Kilmarnock, where it is seen at Bonnyton in March 1960 not long before its withdrawal.

by Leyland. while those used by SMT were bodied by Cowieson. Scottish Transport also used some normal-control Tillings-Stevens coaches in their smart red livery with black roofs (Midland 'Red' are not the only operator to have used this distinctive colour scheme!)

In the early post-war years Western had few coaches available. A few AEC 'Regals' with coachwork by Burlingham and Brush and a batch of Alexander-bodied Leyland PS1's were used. When a fleet of Burlingham bodied Leyland PS1's replaced the old second-hand LT5 and LT5A 'Lions', which had been specially rebuilt and rebodied for the London service, the latter were relegated to various other duties including the Lancashire services. The PS1's and some of the AEC 'Regal' IVs, which replaced them, followed in due course. Later still, some of the Guy 'Arab' LUFs which replaced the AECs were, in turn, demoted. In each case, the luxurious London coach seats were replaced by more austere accommodation. The heavyweight Alexander-bodied Guy 'Arab' UFs, known throughout Western as the 'Bombers', were regular performers. The favourite vehicles for the service nowadays are the unique Bristol MWs with Alexander coachwork and the more recent 36 foot-long Leyland 'Leopards' with Alexander Y-type bodies. Some of the latter are in the black

Western SMT's CA932 at Kilmarnock in 1964, before withdrawal, was originally used on the London service but was frequently seen in Lancashire.

and white livery which is normally reserved for the London coaches, although, the 'Bombers' also wore these attractive colours throughout their working life. These superb vehicles invariably draw many admiring looks from passengers on other coaches, even though many of the Ribble coaches are equally comfortable inside.

Scottish Omnibuses used coaches from their large fleet AEC 'Regals' with bodywork by Duple, Burlingham or Alexander. These were gradually phased out as new Bristol LSs and MWs and AEC 'Reliances' appeared on the scene.

Ribble used many of their old Leyland TS7 and TS8 'Tigers' and the small batch of PS1s with Burlingham coachwork delivered before the advent of the underfloor engine. However, the most frequent performers for many years were the well-known Leyland-bodied 'Royal Tigers'. These, of course, are now in the process of being withdrawn. The Burlingham 'Seagull'-bodied 'Tiger Cubs' which followed them were also used regularly. They too have now been largely replaced by the various batches of 36 foot-long Leyland 'Leopards' which Ribble have placed in service in recent years. The 'Gay Hostess' 'Atlanteans' were tried for a time, but they do not seem to have been successful on these services.

Both North Western and Midland 'Red' use 36 foot-long 'Leopards' for their share of the services. West Yorkshire, when working through, formerly used Bristol Ls of their 'EG' class, but these were replaced by 'EUG' class LSs which are still frequently used despite the advent of newer coaches.

On the day services, there are frequently insufficient coaches or dual-purpose vehicles available to meet the demand at busy periods; then all sorts of single-deck buses are pressed into service. Western has often used units from the small batch of Guy 'Arab' LUFs with Alexander bus bodywork, while Ribble has turned out 44-seater 'Royal Tigers' and 'Olympics', and a batch of second-hand pre-war 'Tigers' with Burlingham bus bodies used to be regular performers.

Even using buses, the operators are unable to find sufficient vehicles for the busiest weekends—well over 100 coaches are required to cover the services in one direction at peak weekends. Vehicles are, therefore, hired from associated companies and independents. Some of the more interesting vehicles turned out in the past have included Alexander's

One of the 'bombers'—KG968 of the Western SMT fleet—at Kilmarnock.

'PC' class 'Royal Tigers' with bodies similar to Western's 'Bombers', and Central's heavy-weight 'Arab' UFs with similar bodywork but with bus seats, including one batch with cut-away rear entrances. Coaches from the fleets of most Lancashire independents have appeared in Scotland carrying "on hire to Ribble" stickers. Blackpool coach operators are particularly frequent hirers; their coaches are used mainly for day tours during the week and are usually free at weekends.

And what of the future? Continued development and improvement of the existing services seems assured. At the time of writing Ribble and the Scottish Bus group, not content to rest upon their laurels, are planning the introduction of new, faster services for

THE LANCASHIRE-SCOTTISH EXPRESS SERVICES

Right. Jackson's FBV808, of Blackpool, carrying the usual "on hire to Ribble" notice as it heads north through Lockerbie.

Below. Central SMT's K54 —a Guy 'Arab' UF with Alexander bus body—was until withdrawal one of an unusual class that was often hired for the day services to Lancashire.

the 1966 season. Application is being made to the Traffic Commissioners to introduce new express services from Glasgow and Edinburgh to Liverpool, Manchester, Blackpool and Morecambe. Except for refreshment breaks, these services will operate non-stop and will complete the journey in about $6\frac{1}{2}$ hours to Morecambe, just over 7 hours to Blackpool, or a little more than 8 hours to Liverpool and Manchester. They will operate on summer Saturdays only and, to forestall possible objections by British Rail, it is proposed that the coaches used should come out of the duplication allowance already authorised on the existing services.

In conclusion, I should like to acknowledge the assistance so readily given by Mr. G. D. Ramsay, Traffic Manager of Western SMT, and Mr. J. Mack, his counterpart with Scottish Omnibuses, in the preparation of this article. Mr. J. W. Tweedie, Traffic Manager of W. Alexander & Sons (Midland) Ltd., supplied details of the original Blackpool tour and even spoke to the driver, Mr. Donald Daly of Stirling, who was in charge on the very first run and who retired in 1963 after fifty years' service with the firm. Fellow enthusiasts Mr. R. L. Grieves and Mr. J. K. D. Blair supplied much of the historical data.

"Nulli Secundus"

The story of the
AEC/LGOC NS buses

by G. J. Robbins

THE AEC NS-type motor bus probably did more than any other type to bridge the gulf between the original, or "Ole Bill", type of open-top bus and the present day covered-top double-deckers. At the time that the first NS appeared on the streets of London in May 1923, all buses were open-topped, had solid tyres and were restricted to a speed of 12 m.p.h., and the drivers were not protected by windscreens. All these were features—perhaps restrictions would be a better term—of the motor bus from its early days. But when the last NS bus ran in London in 1937, all buses had covered tops, were fitted with pneumatic tyres and travelled at speeds in excess of 20 m.p.h., and drivers had been given the protection of a screened cab. There had been other improvements too and the AEC STL-type was fast becoming the standard vehicle in the Metropolis. It seems astounding that such changes could have taken place in a 14-year span, but this account will describe how it happened in London and elsewhere.

The NS-type of chassis was developed jointly by AEC and the London General Omnibus Company, with a low centre of gravity so that the body could be fitted with a top cover; this was an urgent necessity as inclement weather virtually converted open-topped vehicles into single-deckers. So NS1 was built with a top deck cover, although it was a temporary one which could be removed in twenty minutes. The vehicle was subjected to tests early in 1923, and then submitted to the Metropolitan Police for examination; but they would not relax their stringent regulations and allow the covered-top bus to work in London, fearing that it would be top heavy. The capital, therefore, had to wait nearly another three years before being able to sample the advantages of a double-decker with a roof. In the meantime, Birmingham Corporation had successfully produced a top-covered bus of the same style as that proposed by the LGOC, but mounted on the higher 504-type chassis which was the same height as that of the General S-type buses. This vehicle, numbered 101 in the Birmingham fleet and registered OL8100, was the first of several similar buses, and all new Birmingham double-deckers since have had top covers.

Notwithstanding this police discouragement, the LGOC and AEC went ahead with the quantity production of NS-type buses, even though they had to enter service as open-top vehicles. The first twenty or so were put to work on May 10, 1923, on route 11 between Shepherds Bush and Liverpool

Street. These vehicles, which were based at Hammersmith (R) garage, were NS5 onwards, as NS2-4, as well as NS1, were being used for experimental work.

The choice of NS for the type code of this improved design was strange, as up to that time all London General buses bore a single type letter. It was suggested that NS indicated "no step", but the LGOC claimed the letters stood for "nulli secundus" as the design was regarded as "second to none". Contemporary accounts in the trade press, which was most enthusisatic about the new type of vehicle, indicated that the single step from road to platform was only 13 inches high, which must have seemed wonderful at that time although it is now commonplace and standard for most double-deckers; the NS platform was 12 in. lower than that of the S-type, which needed three steps to reach it. At 15 ft. 6 in., the NS wheelbase was six inches longer than the S wheelbase; and the laden weight was 8 tons 7 cwt., compared with 8 tons 10 cwt. of the S, 7 tons for the K and 6 tons for the B-type. The 35 h.p. water-cooled engine had four cylinders, cast in pairs, of 108 mm. bore by 140 mm. stroke. The clutch was of the multi-disc pattern,

These three views—NS1 (*above*), NS13 (*right*) and NS2300 (*opposite page*)—show the rapid progress in passenger and crew amenities between 1923 and 1934; the prototype (NS1) is seen as originally built with top cover for inspection by the police, who did not then approve the inovation; this resulted in the first NSs taking the road as open toppers—NS13 is shown in 1925 when working on suburban route 65A soon after the small route stencils in the lower-deck windows had been replaced by boards on the top-deck sides; NS2300 shows the type in its almost final form and was one of 50 with wider bodies mounted on the later 422 chassis, as can be seen by the ADC radiator.

[*London Transport*

running in oil, and there was a constant mesh gear box with helical gears; the back axle was of the double reduction type having a worm gear in the centre, mounted in the differential and two pinions on the extreme outer ends of the driving shafts, which engaged in two internally toothed drums on the wheels themselves. The constant mesh helical gear box it was claimed, was the most silent of any box and it superseded the change gear box which was apt to be noisy. The large diameter wheel, first introduced with the AEC S-type vehicle, was retained for the NS.

The first NS chassis—NS1—although built late in 1922, did not enter service until July, 1923, as its registration number—XO1019—implies; the earliest registration number applied to an NS—XN1745—was to be found on NS25, and by May 1923 when the first of the type went into public service, nealy 50 NS buses had been built. NS2 was an experimental chassis which, for sometime, carried a lorry body and ran under trade plates; it did not become a bus until March 1924, when it was given the registration number XR1442, and by that time more than 1,100 NSs were in service. The reason for the late appearance of NS3 and 4 seems to have been 'lost', as the chassis were not fully completed until August 1923, and they were registered—XO9273 and 9268 respectively—among the NS500 series.

I recall seeing the new buses, during a trip to London with my father and brother, soon after they first entered service, and I remember being surprised as I had not seen any reference to them in the Press. Our first call was to Westminster, and I was pleased that we had to go to Fleet Street and were thus able to travel on one of the new buses. The comfortable seats of the lower deck were most noticeable as they had upholstered backs instead of the wooden backs in the earlier K and S types. There was more glass in the front bulkhead and I particularly noticed that the driver's head could not be seen from inside the bus; this was due to the fact that the bus body was much lower, in relation to the cab and engine, than in earlier types and necessitated the fitting of a rather attractive white canopy over the driver. The use of the large route stencil, introduced with the S-type some two years before, was continued—but only on the front and back of the bus. As the side ventilating windows on the NS were narrow, the illuminated stencil boxes for route numbers on the sides were quite small, but the use of these was discontinued after the route renumbering in December 1924 and was not resumed until later covered-top days; in the intervening period route numbers were displayed on boards on the upper deck sides in similar fashion to the K-type vehicles. Another change involved the boards at the front which gave details of the streets and places passed en route, as there was less space for the boards above the canopy. This was overcome by using a shallow, but wider, board which was changed from side to side, but the same amount of detail was shown. However this meant there was no room for advertisements on the front.

Between May 1923 and May 1925, the introduction of open-topped NS buses was almost a daily occurrence; over 1,700 were produced during that period and AEC was kept fully occupied. However, they were unable to cope with them all, and more than 500 NS chassis were built at LGOC garages. Most of the bodies were built by Short Bros., of Rochester, but some came from the Brush concern and quite a number were constructed in the General's Chiswick Works; nevertheless, all the bodies were to the same pattern irrespective of the builder.

At first NS buses were allocated to the main central London services, and after route 11, Mortlake garage had them for routes 9 and 73, Tottenham for the 73, Palmers Green for 29 and Seven Kings for route 25; these routes

Upholstered seat backs in the lower deck of the NS were a great improvement; note the higher position of the driver's seat.

[*London Transport*

NS146 working on the short 92 service from Sudbury Town station to Wembley, for the British Empire Exhibition in 1924.

[London Transport

will be easily recognised as they are much the same today. Gradually the new type spread over most of the General system and allowed many pre-war B-type buses to be withdrawn; many NSs replaced the B-type direct and I recall this happening on such routes as 4, 49 and 68 for example. The new vehicles were also most useful in dealing with the extra traffic generated by the British Empire Exhibition held at Wembley during 1924 and 1925. Many of the services extended to Wembley for that event were worked entirely by the NS type, including routes 8, 15A, 18, 43A and 83. A particularly interesting service was the short 92 route from Sudbury Town station to Wembley, which required only two buses in 1924; the only information on the route board was "via Harrow Road", and the destination boards were changed after each journey.

At the time that the first NS was produced it was stated that it would not be available to other operators for some time and, in fact, all NS chassis produced during 1923 and 1924 were taken by the LGOC. The first exceptions were 25 vehicles for the National Omnibus & Transport Co., which introduced them on a number of Country routes north of London on behalf of the General. The vehicles—standard open-toppers—were the property of the General and, even though they carried the fleet name NATIONAL, they were numbered between NS 1024 and 1741.

However, they carried Hertfordshire registration numbers in the NK and RO series. These buses operated mainly from Watford, Hatfield and Hemel Hempstead garages on routes that were at first numbered between N1 and N14, but after December 1924 those routes that entered the Metropolitan Police area became 301, 303, etc., by which numbers most are still recognisable today.

By 1925, open-topped NSs were to be seen on Country routes south of London. Again, these were General owned buses, but were operated by the East Surrey Traction Co. for the LGOC, on a similar basis to the National in the northern area. Some 25 buses were allocated to Reigate and Leatherhead garages for use, in the first instance, on routes 405, 406, 408 and 414. The first East Surrey bus —NS1610, registered as PD3470—was well in advance of the others, no doubt for trial purposes; the remainder were in the late NS1600 and 1700 series and carried PE registration numbers.

A decision to make the NS chassis available to other operators was made at the beginning of 1925 and, from a study of the trade papers of the time, it seems that one of the first to be sold to another concern was a single bus shipped to Argentina for use in Buenos Aires. This was reported in February 1925, and photographs show a covered-top NS being sent overseas some eight months before covered-top buses were to run in London.

By June 1925, buses up to NS1737 had been built for the LGOC, but NS1733 was the last to be put into service at that time; the other four chassis were retained for another

One of the 25 NS vehicles operated by NATIONAL on certain country routes—NS1047 is seen on route 302 in Watford after it had been fitted with 'balloon' tyres.

[W. Noel Jackson

NS1737—one of the first four London buses built with a covered top deck—proceeding to Loughton on route 100B in October 1925.

[London Transport

experiment with covered tops. Short Bros. built four bodies that had more permanent top covers than that in the NS1 experiment. Upholstered seats were provided on the top deck, similar to those inside, together with sliding windows to ensure plenty of ventilation. The four buses, NS1734–1737 (XW 9881–9884), started working on October 2, 1925, out of Loughton garage on route 100 from Epping Town to Elephant & Castle, which is now part of route 20. The experiment, which was well publicised in the daily papers, was a success from the start—it was, of course, the right time of year to introduce it—and I remember breaking my journey home one Saturday especially to enjoy the novelty of a ride on a covered-top bus. It appears that there was some difficulty in obtaining approval for the four buses to run in London, as a *Daily Mail* report, dated September 11, 1925, shows that they were ready then; and the October 1925 issue of *TOT Staff Magazine*, correcting a previous report, stated that the vehicles were to have been licensed before the end of August for the Wood Green–Farningham route, but a hitch had arisen. The experiment continued for about four months and the following extract from a newspaper cutting of early March 1926 makes interesting reading:—

"London is to have 200 covered-top buses. The first four, placed on the Elephant to Epping route, have been so successful that others are being added. Nine more covered buses, our representative was informed yesterday, have now been put on to the streets by the LGOC. They were running this weekend on the route from Shepherds Bush to Liverpool Street. Passengers find them much more comfortable than the old type of bus which is being replaced. There is room for 28 passengers in the padded top deck seats, handrails up the gangways, sliding plate-glass windows, rubber padded—a real luxury vehicle. They do not appear unwieldly and the public are enjoying the experiment of riding in a covered bus, and during a rainy spell, instead of there being a scramble for the inside seats, the competition is to get on top and look at the unfortunate people on the old buses shrinking under waterproof sheets".

At the time that the General ordered these 200 covered-top buses, an order was also placed for 14 new open-top NSs for East Surrey and National; the covered-top vehicles were intended to be NS1739–1938 with bodies by the LGOC or Shorts, and the 14 open-

Route 11 was, in March 1925, the second service to have covered-top buses, and NS2 was among the first vehicles to acquire a covered-top body in place of a open-top one.

[London Transport

toppers NS1939-1952 with Brush-built bodies. However, the open-top bodies were ready first and so they were mounted on the following chassis — NS1739-43/7/50/2/3/6-8, NS1761/4; the first three went to National and the others to East Surrey, and nearly all were in service before any of the LGOC batch. Consequently, when the first few of the covered-top bodies were completed they were mounted on older chassis that had come in for overhaul, and I recall seeing NS2, 84, 1144, 1184, 1236, 1312, 1313 and 1647 carrying top-covered bodies on route 11 at the beginning of March 1926, which were joined by some new ones, such as NS1744, 1745, 1751, etc. When, some weeks later, NS1939-1952 came on the road they had old overhauled open-topped bodies and were allocated to LGOC central routes, most of them working out of Forest Gate or Putney Bridge garages; they were all subsequently fitted with covered tops.

The original 200 covered-top bodies could always be distinguished from subsequent deliveries, as they apparently had a heavier roof with internal struts; later bodies had a sheet metal roof without struts. Routes 11, 29, 76 and 25 were again the first to have the latest type of bus.

AEC had given type numbers 405, 406, 407, 408 and 409 to the NS chassis, but the first four appear to have been variations of those delivered to the General. The 409-type, however, was produced for sale to other operators, and the chassis numbers commenced at 409001. In May 1925, East Surrey purchased eight NSs for use on its own services and, as these were not connected with the General in any way, the vehicles were not numbered in the LGOC's NS series but carried only the East Surrey fleet numbers 160-167; they were registered PE2420-7. Several municipal undertakings bought the 409-type chassis, including Hull, Liverpool, West Bridgford and Warrington. Hull had two vehicles with bodies by Short Bros., which were numbered 14 and 15 and registered KH3428-9; West Bridgford also had two buses, but with Brush bodies. Unfortunately it has not proved possible to trace full details of the sales to provincial operators—I understand South Wales had one—but it seems that most of the buses, if not all, had covered-top bodies almost identical with those used in London.

Another concern that considered the covered-top 409 type a useful vehicle was Waterloo & Crosby Motor Service Ltd., of Liverpool, which bought eight covered-top 52-seat NSs towards the end of 1925 for use on the Seaforth–Crosby route. The bodies were identical with the four then running in London and the Crosby concern was proud of the fact that it possessed eight covered-top buses when London had only four. Two more NSs were added later and all ten vehicles passed to Ribble, with the controlling interest in Waterloo & Crosby, in 1928. Glasgow General Omnibus & Motor Services, of Hamilton, which later became Central SMT Co. Ltd., placed an order for five covered-top 409s with AEC in 1926; these became G1-5 in the fleet—which incidentally had the fleet name "G.O.C."—and were registered VA5706, 5707, 5708, 5743 and 5744.

It is interesting to note that the 409 type was also going overseas and, in February 1926, two were sold to the Equitable Arts Company of Pittsburg, Pennsylvania, U.S.A.; one was an open-topper and the other had a covered top, but both had an offside entrance. A study of the trade Press of the period shows that another London-type covered-top NS bus travelled right across Europe to Budapest in the summer of 1926. The bus carried the registration number RM2561, which seems a strange coincidence in the light of present-day tours by Routemasters! It is not clear if the trip to Budapest was just a demonstration run or whether the bus had been sold; but it certainly had a difficult journey, meeting obstacles in the form of bad roads, steep gradients—sometimes as much as 1 in 3—as well as gateways, arches and overhanging trees. It went by way of Hamburg, Berlin, Leipzig, and Bayreuth to Regensburg, but had to complete the journey to Budapest on a barge down the River Danube.

So far no mention has been made of NS1738. This vehicle was something of a puzzle; it was not a double-deck bus, in fact, it was not a bus at all and the chassis had little in common with the NS chassis, but was almost identical with the AEC 419 chassis then being introduced for coaches; possibly it was given a number in the NS series as it had an NS engine. NS1738, with registration YN3799, had a coach body and was used for private hire work. Another 'oddity' was NS1953, which had a standard covered-topped body mounted on a second-hand chassis—with the strange registration of BT7649—obtained from Hull & District Motor Services.

The years 1926 and 1927 were full of activity so far as London was concerned; new covered-top buses, numbered between NS 1954 and NS 2296 were placed in service and most of the older NSs with open tops had roofs fitted. Route after route was converted to covered-top operation and one did not know which route would be dealt with next; most of the main services in central London were changed over quickly, and then some of the outer suburban routes, but it must be remembered that there were still large numbers of the older K and S types which continued as open-toppers. Before a change-over, each route had to be surveyed and this often meant that overhanging trees had to be cut back.

Another interesting experimental vehicle was NS2050, which was designed to work through the Blackwall Tunnel on route 108—a service operated hitherto by single-deckers. The new double-decker—covered-topped of course— had a slightly narrower body than other NSs, which meant that the normal transverse seating arrangement could not be used. Instead, all the seating was longitudinal; that on the lower deck being arranged each side of a central gangway, while upstairs it was arranged in "knifeboard" fashion with a sunken gangway on each side. The roof had a pronounced dome shape to clear the tunnel wall. NS2050, which had 46 seats, entered service in April 1927 and, six months later was joined by 24 similar buses numbered between NS2210 and NS2239. These 25 vehicles were the first London buses to have enclosed staircases. When they were put into service route 108 was altered to run between Bromley-by-Bow and Forest Hill, later Crystal Palace, as the northern section was unsuitable for double-deckers.

NS2051/2/3 never worked in London as they were sold back to the chassis builders soon after their standard covered-top bodies had been completed, so that AEC could fulfil an order for three buses from Greyhound Motors Ltd., of Bristol. They were registered HU8157/8/9 by Greyhound, which was later taken over by Bristol Tramways, but only one of these NSs was painted in Bristol colours; another subsequently became a race track totalisator. Yet another 'odd man' was NS2231, which the General acquired early in 1927 from a Mr. Antichon, of Yorkshire; originally a single-decker, with registration number WU6715, the LGOC fitted it with a standard covered-top body.

The year 1928 saw the introduction of the second of the main improvements to the NS type, when pneumatic tyres made their appearance. The first London double-decker to be fitted with 'balloon' tyres were the six-wheeled LS of the General and the Guys worked by Public, which appeared in 1927, but such tyres were not permitted on the 4-wheeled NS vehicle until July 1928 after certain regulations had been amended. The difficulty was that 4-wheeled buses required larger rear tyres, which made the NS type wider than the statutory width. The first pneumatic-tyred NS introduced by the General had several improvements compared with earlier varieties; it was mounted on the later ADC 422-type chassis and had a slightly wider body and more luxurious seating. The windows on both decks were controlled by individual handles and worked up and down—a great improvement on the horizontal sliding windows of earlier covered-top vehicles. An order for 75 of these new buses was placed by the General but only 50 numbered NS2297-2346, were delivered. In May 1929 a further six of this style were

The only single-decker in the London NS fleet was this private hire coach—NS1738 (YN3799)—which had little in common with the remainder of the type, except possibly the engine.

[London Transport

The advent of the NS, with its protective roof and lower body, enabled double-deckers to operate through Blackwall Tunnel on route 108; even so, the special bodies had longitudinal seats in the lower saloon (*top left*), and in the upper saloon (*left*) 'knifeboard' seats were used in some vehicles and 'bucket' seats in others; the enclosed staircase was another feature.

[*D. W. K. Jones; London Transport*]

The same chassis, but different bodies—WU6715 in the service of Mr. Antichon (*above*) and in London Transport days as NS2231 (*right*).
[*Tramway & Railway World; D. W. K. Jones*]

introduced and these carried the numbers NS2372-2377, leaving the numbers between 2346 and 2372 unused. These new buses were sent to Cricklewood garage for use on route 16A (Cricklewood–Victoria) and also to Mortlake for route 33 between Richmond and Aldwych. Gradually during the next five years a large proportion of the NS fleet was fitted with pneumatic tyres but many were not changed as the NS bus was being replaced by newer types.

One particular advantage of the pneumatic-tyred bus was that its speed limit was 20 m.p.h., compared with the 12 m.p.h. of the solid-tyred vehicle, and so pneumatic-tyred buses were to be found on some outer suburban and country routes which could be speeded up; the open-topped NSs operated by National and East Surrey were all converted to 'pneumatics' during 1928.

One of the companies operating in London and in association with the LGOC was the British Automobile Traction Co. Ltd., which used the fleet name BRITISH and which operated 33 buses from a garage in Camden Town on route 24 (Hampstead Heath–Pimlico) and also on Sundays on route 63 (Hampstead Heath–Honor Oak). A fleet of Daimler buses had been used for many years, but during the latter part of 1927 NSs on 409-type chassis with 48-seat covered-top bodies took over. They were not numbered in the NS series, of course, but carried the BAT fleet numbers 501-533 on their bonnets; when London Transport acquired them in July 1933, they were renumbered NS2379-2411. They also lost their dark green British livery in favour of the ubiquitous red, but none was ever fitted with pneumatic tyres and all were withdrawn during 1934.

Some NS-type buses could be seen in London carrying the METROPOLITAN fleet name; these were vehicles in a separate fleet that were operated by the LGOC on certain selected routes from General garages on behalf of Tramways (M.E.T.) Omnibus Co. Ltd. In 1927 the MET fleet consisted of K and S-type buses but, in December that year, 21 open-topped NSs replaced a similar number of S-type buses on route 88A (Acton Green–Mitcham), which was being diverted to work to Grove Park, Chiswick, and meant running under a low bridge for which the lower NS-type was more suitable. The altered routeing did not last long, but the NS vehicles remained in the Metropolitan fleet and many later received top covers. Early the next year, the MET fleet was increased by the transfer of 70 more NSs from the General; these were all covered-top vehicles which were then working on routes 8 and 160, between Old Ford and London Bridge or Willesden, out of Willesden or Dalston—later Clay Hall—garages.

Not all the original open-topped NSs were given top covers—in 1931, 70 open-toppers remained in use on a few routes with low bridges such as 116 and 120. However, five of them were transferred to East Surrey, with the operation of route 70 (Morden station–Dorking), and the remainder drafted to Old Kent Road garage which, at that time, could not accommodate full-height buses in part of the garage. Strangely, the number of open-topped NSs increased during 1932 and 1933, when fifty covered-top buses were 'decapitated'; some were used on normal services, but most were licensed only for private hire work—mainly for operating to the races at Epsom Downs where they became ready-made grandstands. Another 20 open-toppers were provided in the General fleet in 1932 when some of the former East Surrey and National buses exchanged bodies with covered-top LGOC vehicles; this move also resulted in covered-top NSs working on country routes for the first time. The difficulty with the low roof in Old Kent Road garage was overcome in 1933, so all open-toppers were withdrawn from regular service.

The third main development with the NS—a glazed cab to protect the driver—started in 1929. The first experiment in this direction

occurred in March with NS198, but it does not appear to have been very successful. A modified form of apron windscreen was tried on three more NSs in August and September that year but, after a short time, these were removed leaving the three vehicles with larger than usual driver's cabs. It was some two years before a more satisfactory style of glazed driver's shield was devised and fitted to all NS-type buses.

Other interesting developments included the fitting of a sliding roof to the top deck of NS397 in 1930—I have no idea how long this unusual innovation lasted—and experiments with 6-cylinder engines; NS1758, an East Surrey open-topper based at Reigate, and NS2015, a top-covered vehicle on route 60 (Cricklewood–Old Ford), were both fitted with larger engines. Quite soon after the covered-top bus was introduced to London, the front route number stencil was moved from the time honoured position in the centre of the driver's canopy to a position above the front destination board and between the upper-deck windows; but when the later type of covered top, with drop windows, was introduced in 1928 the front route number was positioned on the cab roof. In this connection it should be mentioned that as well as the 50 new buses with the new type of top cover, many older vehicles that were fitted with covered tops after 1928 also had the improved type of top deck with drop windows; there were some 250 spread throughout the NS fleet.

The last NS-type bus to enter service in central London was NS2290, which appeared in February 1930. The chassis had been used at Chiswick for training purposes in the driving school, and when replaced by an ST chassis—then being introduced—this NS was given a covered-top body and the late registration number GC3953, which was in the middle of a batch of ST registration numbers. It is unfortunate that no illustra-

British Automobile Traction (opposite) operated 30 NS-type buses on route 24 from Hampstead Heath to Pimlico; twenty-one open-top NSs were transferred to the **Metropolitan** fleet and subsequently used on route 88A, which terminated at The Cricketers, Mitcham (*right*)—this view shows two of the type with, to the left, an S-type on route 105; some of the NSs in the London **General Country Services** fleet, previously operated by National, acquired covered-top bodies in 1932—NS1616 (*above*) is seen in Hemel Hempstead on route N14, which had retained its N prefix as it did not enter the Metropolitan Police area.
[*W. Noel Jackson; London Transport; J. F. Higham*]

NS2154 (left) in London Transport days represents the type in its final form after fitting with pneumatic tyres, the improved style of covered top deck and a cab windscreen. **Oxford** 106 (below) and Newcastle-upon-Tyne 89 (bottom) were among 23 NSs, with Chiswick-built bodies on the 422 chassis, which went to the provinces.

[*London Transport*
L. L. Cooper;
D. W. K. Jones

tions seem to have survived—if any were taken—of this bus or of some of the other unusual vehicles, such as NS 1953 and NS 2015.

Mention has been made of the improved NS chassis, designated 422 by AEC, which were built between 1927 and 1929. In addition to the 56 buses with this type of chassis that were operated by the General, one hundred or so were sold to other operators. The first seventy 422 chassis which had aluminium wheel centres, made up a large order for Anglo-Argentine Tramways. One other chassis went overseas—to Walford Transport, of Calcutta—but the others remained in Britain. Newcastle-upon-Tyne obtained six in 1927/28, which became 85-90 in the Corporation fleet and were registered TN6519-6524. Two NSs went to Warrington Corporation and ten to the City of Oxford Tramways; those in the latter batch were given the fleet numbers 104-110 and 1, 7 and 9, their registrations being WL5337, 5338, 5346, 5347, 5352, 5362, 5363 5698, 5987, and 6165. The Glasgow General Omnibus Co. placed a repeat order for NSs in 1927, and this time had five of the 422 variety which became G6-10 (VA6941-6945). The 23 buses for these four operators all had 54-seat covered-top bodies built by the LGOC at Chiswick. The last purchaser of the NS-type was Derby Corporation, which had six in 1929; they had 48-seat Brush bodies with enclosed staircase and were numbered 17-22 (CH7885, 7886, 8337-8340). The only 422-type chassis to carry an open-top body was a demonstrator, which was registered MP1460; it was, in 1929, added to the East Surrey fleet, but was not numbered in the NS series until London Transport days when it became NS2378.

In the first half of 1933, six former East Surrey vehicles passed to the LGOC when it took over the operation of the route between Sidcup and Orpington—then numbered 411, but now 51. Three of these buses already had NS numbers, but the others were three of the eight vehicles owned by East Surrey which, it will be recalled, had not carried NS numbers; the General, therefore, issued to them NS2347-2349, which had not been previously used.

The withdrawal of the NS type started in 1932 and nearly 200 had been taken out of service by the time London Transport became responsible for operations on July 1, 1933. Naturally, it was the solid-tyred vehicles that

Three of the eight buses purchased direct by **East Surrey** were transferred to the General fleet early in 1933 and were then given NS fleet numbers —NS2347 is shown still at work on former East Surrey route 411.

[J. F. Higham

A few standard covered-top NSs with solid tyres including NS1849, were retained for use in the narrow Rotherhithe Tunnel for some two years after other solid-tyred buses had been withdrawn, because of tyre-rubbing on the tunnel kerbs.

[London Transport

were disposed of first and, by July 1935, there were only 34 of that variety still in service in London. But these continued in use until 1937, owing their retention to the special conditions pervailing in the narrow Rotherhithe and Blackwall tunnels. Blackwall Tunnel, served by route 108, required the specially-designed vehicles that have already been described, but those on route 82, through the Rotherhithe Tunnel, were standard covered-topped NSs except for the solid tyres. The solid tyres were retained as it was considered there would be excessive wear on pneumatics through the continuous rubbing of the tyre sides on the nearside kerb in the tunnels. Although one of the special tunnel buses—NS2213—was fitted with pneumatic tyres in 1935 to see how they would stand up to the extra wear, none of the others was similarly treated.

The year 1937 saw the demise of the NS-type in London, at least in public service, for during that year they were withdrawn quite rapidly in favour of new STLs and, by October, only 70 were still in use on four routes—4 (Bermondsey-Finsbury Park), 166 (London Bridge-Aldwych), 178 (Croydon-Addiscombe) and 197 (Croydon-Norwood Junction). It was to route 166 that the doubtful honour fell of bringing to a close another chapter in London's bus history; the end came on the night of November 30, 1937, and, even though it was a wet wintry evening,

Several enthusiasts were among those making the farewell trip on NS1974 when it set out from London Bridge on the final NS journey on a wet night in November 1937.

[*London Transport*

Some NS buses had a further lease of life as departmental vehicles—NS1961 (*above*) became a tramway overhead repair lorry and NS760 (*below*) acquired a quite modern looking body for use as a trolleybus wire lubricator unit.

[*D. W. K. Jones*

the final trip from London Bridge to Aldwych by NS1974, of the now closed West Green garage, was suitably celebrated and recorded for posterity.

However, as has often been the case, a number of buses found a 'new life' in other spheres; thirteen NSs were sold to a Mr. Middlemiss, of Upminster, who used them to take poor East End children on outings to the London countryside, and London Transport retained some 25 for use in other capacities. The Tramways Department had nine, which were fitted with special bodies and used as tower wagons, and four had their bodies adapted for use as tree-cutting vehicles. The other 12 saw another 10 to 14 years' service as staff canteens, with the lower deck fitted out as the kitchen/servery and the upper deck as a 'restaurant'. These mobile canteens were at first painted green, but later red.

I have no record of any other NS-type vehicles having remained in use after withdrawal from service in London, and the only known survivor is the specimen in the Museum of British Transport at Clapham. This is NS1995 (YR3844), a standard covered-top double-decker with pneumatic tyres, which was introduced in February 1927 and, therefore, had ten years' active service in London, but has now been in honourable retirement for almost 30 years. However, it remains to remind us of a most interesting type of motorbus, which—in the 'twenties at least—certainly was in many respects "nulli secondus".

The Postal Buses of Switzerland

by J. Graeme Bruce

IN 1849 the administration of postal services in Switzerland was taken over by the Federal Government, and the separate postal organisations in the cantons, with various postage rates, and intercantonal tolls were suppressed. These early postal services had relied to a large extent on the many stage coaches that operated in the country, although these were mainly local in character but it was only natural that the central government should have some say in the arrangements for transporting mail. However, the federal authority only regulated the principles by which the stage coaches operated, leaving the details of the services to be decided by the needs of the various localities under the control of the appropriate cantonal authorities.

Above. With the Rhone glacier in the background, a 36-seat Saurer coach of the Swiss Postal Bus services descends from the Grimsel Pass.

all photos courtesy Swiss Postal Authority

For 64 years, from 1850 to 1914, the stage coach services were under constant development, until the total kilometres operated exceeded 10 million a year on routes totalling over 7,000 km., and requiring more than 2,200 horse-drawn carriages. In winter many of the operations were maintained, of course, by the use of horse-drawn sledges.

The first motor bus in Switzerland began operating in 1903 between Porrentruy and Damvant, close to the French border in the less mountainous part of the country. This service was provided by a private operator, and the success of the venture, together with the development of the internal combustion engine—especially in France—led the Swiss authorities to consider speeding up the carriage of mails by the use of mechanically propelled vehicles.

In 1904, the postal authorities introduced a mechanically-propelled van in Zurich and in the following year obtained two more.

Progress was, however, slow in that it was not until 1918 that such vehicles were used in Berne and Geneva; in fact, the early days were not altogether successful, as the mechanical troubles were numerous, and there was talk of reversion to horse transport.

Franz Brozincevic, who began as a motor vehicle repairer in Zurich in 1905 and went on to found the FBW concern in Wetzikon, offered in 1908 to operate as a factor and to hire vehicles as required to the post office, on lines similar to those of the McNamara organisation in Britain. The first commercial vehicle built under his guidance, which was known as the Franz-Lastwagen, was assembled at Zurich in 1908, but it was not until 1922 that regular manufacturing began at Wetzikon.

Provided gradients were not too severe, the use of mechanically propelled vehicles showed promise and the postal authorities on June 1, 1906 started operations with passenger carrying motor vehicles on the route between Berne and Detligen. Three journeys a day were operated, the time taken being an hour and a quarter; before the advent of the motorbus this 17-km. journey took two hours and 10 minutes, and the journey time today is 40 minutes.

Three vehicles from three different manufacturers were obtained for this service. The Saurer and Berna were similar in design, but the Martini had an underfloor engine placed under the totally enclosed cab and, for some reason associated with its shape, was nicknamed "Granny". This vehicle could achieve a speed of 21 m.p.h., which was amazingly high for a commercial vehicle of those early days. The engines of all three vehicles had four cylinders and developed 30 to 35 h.p.

Adolph von Martini built the first motor car bearing his name in 1897; it had an engine at the rear. In 1903 premises were acquired at St. Blaise, near Neuchatel, where the manufacturer of commercial vehicles began, but the concern thus established did not survive the world recession of the early 'thirties and ceased manufacture in 1932.

The first Saurer vehicle materialised when Adolph Saurer, already producing stationary engines of various types, decided, in 1898, to equip a passenger carrying vehicle with an existing engine suitable for this duty; production of commercial vehicles began four

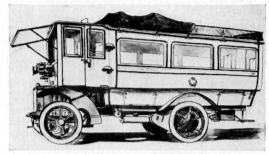

A drawing of 'Granny'—one of the three original Post Office motor buses introduced in 1906.

years later, and the firm of Aktiengesellschaft Adolph Saurer, of Arbon on the shores of Lake Constance, has been in the commercial vehicle business ever since. The company played a big part in the development of the automotive diesel as Saurer worked with Dr. Rudolf Diesel on such a project. The 1903 model Saurer had a leather cone clutch 3-speed gearbox with cardan shaft bevel-gear drive to the rear axle. The handbrake worked on the rear wheels while a footbrake operated on the cardan shaft. Each of the four cylinders of the engine was contained in a separate block with valves arranged at each side operated by separate camshafts.

Joseph Wyss built his first motor car in Berne in 1902 and christened it the Berna. He later formed the commercial vehicle manufacturing concern bearing this name at Olten. Berna vehicles are still manufactured today, although there is now close collaboration with the Saurer organisation dating from the financial difficulties experienced in the depression years of the early 'thirties.

The three original buses were equipped with solid tyres and four-cylinder engines. Accommodation for 14 passengers was provided inside on two longitudinal seats with plush covers, mail and baggage being stowed on the roof. The interior was lit by candle lanterns but there was no heating; the driver, however, was protected from the weather in a totally enclosed cab. Each bus always carried a second man, who normally looked after the baggage and mail but who also took care of the passengers when breakdowns occurred. A second route, from Berne to Papiermuhle, reverted to horse traction for a time, but motor transport continued to serve the postal department although little expansion took place for ten years

For postal administration purposes the whole of Switzerland is divided into eleven postal districts and the postal authority's motor transport section has four divisions, which broadly cover operating, engineering, purchasing and commercial matters respectively. Apart from its responsibilities within the postal department, the motor transport section is also responsible for licensing other operators and issuing concessions, as well as the control of vehicle standards and inspection matters which, in Great Britain, are vested in the Ministry of Transport.

Four different types of bus service operate in Switzerland, three under the direct control of the postal authority and the fourth is governed by the authority's regulations. First, there are the services designated on the "Indicateur Officiel" by the letters PR which are operated by the Postal Authority itself with its own vehicles. The second type covers services operated on behalf of the Postal Authority by agents, and designated PA in the timetables; the vehicles for these services are usually of the same type and style as the postal buses, but are generally painted in the agent's livery even though they may, in fact, be owned and maintained by the Postal Authority. Thirdly, there are the routes to remote villages over roads not suitable for ordinary buses, which are served by large motor cars carrying up to eight people; these are known as "messenger", or PM, services, and are also operated by agents. Finally, routes operated under concessions, which are designated C in timetables, include all the services of the large municipal authorities and a number of private operators; the liveries are quite distinct from the postal buses and, although mails are carried on some routes, the day-to-day operations are not controlled by the postal authorities.

The necessity for this overall type of organisation came about with the advent of the motor vehicle which made a fundamental change in the type of service provided, and the operations, in general, were no longer local in character. The change began in 1916, and one of the Postal Authority routes introduced at that time was that between Locarno and Brissago, on the shores of Lake Maggiore near the Italian border. This service is now operated by Societa Ferrovie Autolinee Regionali Ticinesi, the Locarno concern, under concession arrangements that include the carriage of mails.

In 1918, a route in the St. Gallen postal division between Nesslau and Wildhaus, with an extension to Buchs, was begun with 45 h.p. Saurer vehicles which could carry 16 passengers. Vehicles of this type were capable of hauling a passenger trailer, with seats for 25 more people, in addition to a two-wheeled trailer that could carry a further 1,500 kilograms of luggage. The haulage of trailers in this way was established as a regular feature in 1919. Although there are not many routes suitable for passenger trailers, they are still a feature of this route as well as of a number of others where the roads are wide and steep hills are not encountered and where the traffic demands a greater carrying capacity at certain times. However, luggage trailers attached to passenger vehicles are now a feature on all routes, including those on narrow mountain roads.

In the summer of 1919 the horse stage coach service from Brigue across the Simplon Pass to Gondo on the Italian frontier was replaced by the first of the 'Cars Alpins'. This was a Saurer vehicle having a four-cylinder 35 h.p. engine, which was made up of two blocks with two cylinders in each block. The vehicle had four forward gears and one reverse, and the drive was through a cardan shaft to the rear axle avoiding the use of chains which caused excessive noise on mountain roads. The footbrake was applied to the cardan shaft while the handbrake worked through a beam to the rear wheels. The wheels were provided with solid rubber tyres.

So successful were these operations with motor transport that one hundred army chassis were obtained on loan for conversion to passenger carrying vehicles. Again, three types were provided; the 45 h.p. Saurer vehicles were converted to 'Cars Alpins', with 16-seat charabanc type bodies that had "cape cart" folding hoods and mica side screens for use in inclement weather. The Berna vehicles, with engines developing 40 h.p., became 16-seat buses with luggage racks on the roofs, while the smaller 32 h.p. Martini chassis were given 10-seater bus bodies.

Development of the services then proceeded apace. In April 1919 the first motor service in the Grisons was inaugurated with a route

One of the hundred 16-seat vehicles with "cape cart" hoods and mica windows, that first appeared in 1920 on the Saurer military chassis.

between Reichanau and Waldhaus, south of Chur; the service over the San Bernadino Pass was motorised in 1920, followed in 1921 by those through the Julier and the Grimsel. The 80 km. route across the Julier Pass, in the Grisons, from Chur to St. Moritz was for a time the longest service operated, and it is still one of the longer routes. During 1922 and 1923, motor services were introduced over most of the remaining Alpine passes, although the Fluela and Susten Passes did not have a post bus service until 1946. By the end of 1922 some 600 horses had been displaced and the fleet of motor vehicles had reached the 150 mark, including 62 'Car Alpins' charabancs, mainly Saurers, 61 large buses for 16 or 20 passengers, on Berna or Saurer chassis, and eleven 10-seater Martinis.

The year 1922 also saw several developments in other respects; electric lighting superseded acetylene lamps, pneumatic tyres made their first appearance and, as already mentioned, the first FBW 'Car Alpins' from the works at Wetzikon was delivered. The following year saw the development by Martini of a special 8-passenger alpine car for use along the smaller, remote valleys of Wallis, Tessin and the Grisons. This type of service is no longer operated by the Postal Authority, and all routes which can only support small vehicles of this type are maintained by agents operating "messenger" services.

In 1926 came the first of the so-called all-weather type 'Cars Alpins', still with a fabric roof but with glass side lights. The 6-cylinder petrol engine made its appearance in 1931, to be superseded in 1933 by the Saurer direct

The 17-seater petrol-engined Saurer from which the types I, II and III Alpine motor coach were evolved—the 1929 version (above), and (opposite) the 1932 variety with electrically operated sunshine roof.

injection diesel engine when 14 vehicles entered service for the Post Office. In fact, a 4-cylinder diesel engine had been built by Safir, of Zurich, in 1908; this was designed by Saurer under the guidance of Dr. Rudolf Diesel for commercial vehicles work and, although not successful, did give the Saurer designers experience in this field leading to the production of the successful direct injection engine some 20 years later.

Some 6-wheeled types of all-weather 'Cars Alpins' made their appearance in the early 'thirties, but no more were built for the Swiss postal authorities after 1953. At about the same time, 4-wheeled buses, seating 40 people, began to replace the 16-seater types developed at the end of the first World War.

The outbreak of the second World War in 1939 caused severe curtailment of Swiss bus services as well as the requisitioning of vehicles and other restrictions. The six international services then operating were withdrawn immediately, but they were restored at the end of hostilities. These routes are:— (1) Scuol-Landeck, working between Engadine and Austria jointly with the Austrian Post Office; (2) and (3) the two services over the pass of Fuorn into the Italian Dolomites and to Bolzano, both operated jointly with Societa Automobilistica Dolomite; (4) Lugano-St. Moritz, from the Swiss Ticino to the Swiss Grisons through Italian territory but operated entirely by Swiss vehicles; (5) Brigue-Iselle, over the Simplon Pass and extended to Domodossola in Italy, which is also a solely Swiss operation; and (6) St. Moritz-Munich, jointly with the German Post Office.

Swiss postal buses today penetrate into the heart of the Alps, covering 9,600 km of route —a greater route mileage than that of Swiss railways—and serve 2,620 localities. For the past ten years all services have been motorised, the last horse-drawn stage coach service—between Innerferrera and Cresta (Avers) in the Grisons—having been withdrawn on April 30, 1957, since when the service has been incorporated in a longer route covered by 30-seater vehicles of an agent.

In 1955 the services directly under the control of the postal authorities covered 6,556 kilometres of route and required 476 vehicles owned and operated by the Postal Authority itself and 746 vehicles operated by agents. Together they carried over 22 million passengers a year on 506 routes, 119 operated by the Postal Authority and 387 by agents. By 1962 there was a total of 537 routes operated under the control of the Post Office averaging 58,000 bus kilometres per day and the number of passengers carried had risen to 29 million. There is hardly a road of any consequence now that is not traversed by a postal bus.

Competition, however, is not accepted; only one operator covers one particular route; existing railways—be they local or main line—are adequately protected. Buses serving remote villages are not allowed to run into the central town if a railway operates in the main valley, the bus service operating only to the nearest railway station in con-

nection with the trains, so that, apart from a number of scenic services which have tourist attraction, the bus services are generally operated as feeders to the railway.

Vehicles used by the postal authorities are designed for long life and they are carefully maintained with this in view. There is a central workshop near Berne dealing with overhauls and general engineering work, which was brought into use in 1941, and 58 garages located conveniently throughout the country cover the daily maintenance requirements.

The vehicles in the modern fleet are of six basic types, a very large proportion being of the normal-control layout as the Saurer designs remained very orthodox in this respect for a number of years after the second World War. But as the total number of buses in the country does not exceed 4,000 vehicles, the retention of proven basic designs is understandable.

At the end of the war it became necessary to overcome the shortage of vehicles and, in 1947, a number of 29-seat Alfa Romeo underfloor-engined buses were imported, following trials with a 33-seat prototype of that make.

The first large forward-control vehicle also appeared in 1947; this was a 41-seater from the Saurer stable which was capable of hauling a trailer for a further 28 passengers. Vehicles of this type were tried on the Chur-Lenzerheide service on the lower slopes of the Julier pass, but they were better employed on the Nesslau-Wildhaus, Baden-Endingen and other routes in the lowland areas of Switzerland. The C51U-type FBW vehicles, with underfloor engine and seating 41 passengers, first appeared in the fleet in 1951; with a maximum speed of 65 k.p.h., they now form the backbone of the regular coach services. The trailers which they tow were built by Fritz Moser, of Berne, and accommodate 35 passengers.

The main portion of the fleet, however, consists of 'Cars Alpins', of which there are four types—defined by carrying capacity—in regular use. The smallest—type I—seats 21 people and is usually based on a normal-control Berna chassis with a 68 h.p. engine giving a cruising speed of 55 k.p.h.; some have been in service since 1950 and, although the Postal Authority uses a few on the less frequented routes, the majority are operated by agents over narrow roads. The type II variant, for 25 passengers, is very similar to the type I but is slightly longer and wider; both Saurer and Berna normal-control chassis with 68 h.p. engines, are used, but the bodies have folding roofs. The first vehicles of this type were built in 1939, but a modern version appeared in 1951. Twenty-nine-seat type III vehicles are now the most common of the 'Cars Alpins'. Three manufacturers—Berna FBW and Saurer—have provided normal-control chassis for these buses, which have 125 h.p. engines giving a cruising speed of 74 k.p.h. Two types of body are now in use, one with a folding roof and the other with a fixed head but having sunshine or daylight type roof lights. Finally, the type IV is a forward-control underfloor-engine vehicle, seating 36 passengers, based on either FBW or a Saurer chassis; the prototype appeared in 1954 and engine power is 150 h.p., giving a cruising speed of 79 h.p.h.

A diesel-engined FBW 29-seat 'Car Alpins' type III of 1935.

Passenger carrying vehicles of the Postal Authority all appear in a pleasant yellow livery which can easily be seen on the difficult mountain roads. They do not carry fleet numbers, the licence plate number with the prefix letter P, which is exclusive to Postal Authority vehicles, is sufficient. Unlike Britain, the trailer, whether it be passenger carrying or for luggage, carries a different number. A vehicle hauling a trailer must carry a disc displaying a white triangle on the front of the vehicle; in addition, vehicles on regular post bus services carry a distinctive yellow post horn on a similar disc. Vehicles carrying this post horn disc have a legal right of way and, on specified mountain roads, have to be given preference of movement.

Agents' vehicles, unless hired from the postal authorities, are registered in their own canton and carry the normal cantonal licence letter and number; under certain circumstances, they also carry the post horn disc.

Drivers of Postal Authority buses must be able to speak two languages and, before actually driving in passenger service, have to spend at least a year in the workshops and a further 12 months as van drivers. Twenty-two years of age is the minimum recruiting age for coach drivers and in order to provide the additional summer services required, there are a number of mechanic drivers, who carry out the dual job of overhauling vehicles in the winter and coach driving in the summer season.

An FBW 36-seat type IVu, following the shore of Lake Lugano, with Mount San Salvatore behind, on the international service between Lugano and St. Moritz which passes through Italian territory en route; note the post horn disc on the front of the vehicle.

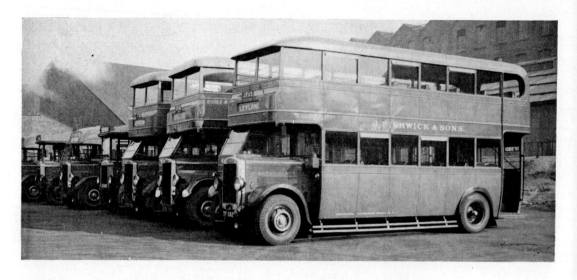

The Leyland Story: 1926-1942

by A. Alan Townsin

This article tells the story of Leyland buses and coaches from the beginnings of the 'Lion' family to the "unfrozen" TD7 models allocated to operators in the dark wartime days of 1942. It is based on articles that appeared in *Buses Illustrated* for October - December 1954 and January - March 1955.

LATE in 1925, Leyland Motors Ltd. of Leyland, near Preston, Lancashire, introduced a new range of passenger models which were designed entirely as such, instead of being, as had been usual until then, based on goods designs to a greater or lesser degree. Five models were introduced, and for the first time the company gave each a name. They were the 'Leviathan' double-decker, the 'Leopard' 38-seater, the 'Lion' 32-seater, the 'Lioness' 26-seater and the 'Leveret' 20-seater. The three larger chassis were of the forward-control type and the two smaller models were of the normal-control or "bonneted" layout. All had four-cylinder overhead-valve engines.

It is important to distinguish the original 'Lion' and 'Lioness' models from their successors of the same names. Designated LSC1 and LC1 respectively at first—the more familiar PLSC and PLC type letters not being used until about the end of 1926—both types had several main components in common. The 'Lion' engine had bore and stroke dimensions of $4\frac{1}{4}$ in. by $5\frac{1}{2}$ in., giving a capacity of 5·1 litres, and most PLC 'Lionesses' had the same unit, but as originally announced the engine of the latter had a bore of $3\frac{3}{4}$ in., making the capacity 3·96 litres.

The transmission was generally orthodox, with plate clutch and four speed sliding-mesh gearbox, but the 'Lion' was then unusual among passenger chassis in having a double reduction rear axle, consisting of helical gears and a spiral bevel unit. It was this which gave the PLSC its characteristic sound when

J. Fishwick & Sons, of Leyland, Lancs., have naturally always been regular customers of Leyland Motors Ltd. The 1946 view (above) shows *Titan*, *Tiger* and *Lion* buses dating from the 'twenties and 'thirties; the nearest vehicle was TF5921—a Titan TD1 with Leyland body, of 1931—and beyond are two TD3 or TD4 models, a lion PLSC, a Tiger TS8 and another PLSC.

[Leyland Motors

THE LEYLAND STORY: 1926-1942

The *Lion 6LSC* helped to build Leyland's reputation 40 years ago; this 1927 PLSC1—TY3673—photographed in the 'thirties when owned by the Lincolnshire Road Car Co. had earlier been in the United Automobile Services fleet, and was originally supplied to A. Proud of Choppington, Northumberland; the 31-seat body is representative of the standard Leyland design for this chassis.

[W. J. Haynes

The *Lioness 6LC1* had quite different styling from the corresponding Lion PLSC, and generally resembled a large contemporary private car; Birch Bros: well-known London-Bedford service was inaugurated with this type of 24-seat coach in 1928.

[Birch Bros.

on the move—an unmistakable whine, not unlike gearbox noise, but which was evident regardless of which ratio was engaged. However it was neither unpleasant in pitch nor unduly excessive and the PLSC was a smooth running vehicle, well above average in this for its period.

The PLC 'Lioness' had a worm driven rear axle and differed considerably from the PLSC in general appearance and layout, quite apart from the driving position. The frame was different, with slightly curved front dumb irons, and the radiator, of quite different shape, was placed almost directly over the front axle beam, as on private cars of the period, rather than further forward. The general impression given was, in fact, rather that of a large private car.

A Leyland-built body was available on the 'Lion 'chassis, and was fitted to a large proportion of those built. Its upright, but well proportioned lines were matched by solid construction which proved to be as trouble free and long lasting—in quite a large number of cases remaining in service for well over twenty years—as the chassis. Leyland-built bodies were, in those days, of composite, i.e. wooden framed, construction. Alternative entrance positions were available and late in 1926 a version on a longer chassis, designated PLSC3, which could seat up to 35, was also introduced. The PLSC3 chassis was built in large numbers and few PLSC1 models were turned out after 1927.

The total number of 'Lion' and 'Lioness' models of this series was over 2,900, and the chassis numbers ran from 45000 to 47960; the series also including the very few 'Leopard' models of this series built. The 'Lion' chassis were, of course, by far the most numerous and examples were to be found in almost every part of the country by the time production of PLSC models ceased early in 1929.

At the 1927 Commercial Motor Show Leyland Motors Ltd. introduced the original versions of the Leyland 'Tiger' and 'Titan', which were to have a considerable influence on British p.s.v. design. Although in many respects well ahead of contemporary practice, they were, as usual with successful designs, based on correct applications of advanced, but sound, principles, rather than revolutionary in conception as a whole.

Most of the main features—a six-cylinder engine of relatively high power, low frame height and vacuum brakes—were to be found singly on other makes of chassis. But their combination together with a large number of detail improvements, many of which have since become normal practice, into a single vehicle provided a new standard of design. This was especially true of the 'Titan' double-decker, which was designed as a complete vehicle—in which the body layout played an important part—with the idea of capturing and extending the double-deck market.

In 1927 large fleets of double-deckers were almost unknown outside London and Birmingham. Most other cities had very limited numbers in service and some of the smaller municipalities had begun to build up fleets, often with six wheelers. Company-owned fleets of double-deckers were comparatively rare, and the idea of running them on inter-urban services was almost unheard of. Generally speaking, the tram reigned supreme in the town and single-deck buses held a virtual monopoly elsewhere. The lack of suitable pneumatic tyres for heavy duty largely accounted for this, and the vogue in six-wheelers which started about 1926 was partly due to the fact that such chassis offered the first practical means of running double-deckers on pneumatic tyres. The Leyland 'Titan' began the reversal of this trend by showing that the complication and heavy tyre wear of the six-wheeler was unnecessary for a normal double-decker.

The standard combination of the original 'Titan' chassis, designated TD1, with Leyland body also constituted the first lowbridge bus as we know it nowadays. The idea of having a sunken gangway at the side of the upper deck was not new, but the single gangway layout and the low-built chassis made the type a practical proposition.

The low-height double-decker has come in for much adverse criticism in recent years, but the side gangway layout allowed the advantages of double-deck operation to be given to routes with limited headroom under bridges. While the centre gangway normal-height double-decker is more convenient and is preferable wherever it can be run, it must be remembered that the travelling public in many parts of the country were not accustomed to double-deckers until the early 'Titans' appeared, and the low build helped to inspire confidence.

The six-cylinder engine was of 6·8 litres capacity, with bore and stroke of 4 in. by $5\frac{1}{2}$ in. It had overhead valves, but unlike the PLSC, where push-rods were used, they were operated by an overhead camshaft, an arrangement generally associated with high

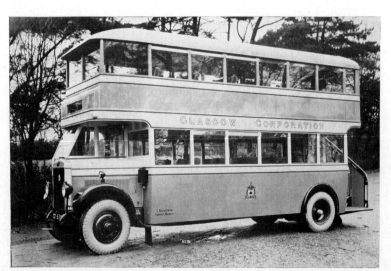

Glasgow Corporation was one of Leyland's last customers for the *Titan TD1* model; Glasgow 51, GD9701, delivered early in 1928, was the first vehicle supplied, and is representative of early Titans—its chassis number was 70021. This form of radiator was used for the first year or so, and the open-staircase body was in the early style of the standard Leyland product.

[*Leyland Motors*

performance private cars. In this respect the T-type engine, as the TD1 unit was called, resembled that in the famous Leyland Straight Eight car of 1922, although the camshaft drive was new. The T-type engine formed the basis of a whole range of Leyland petrol and oil engines, some of which remained in production until 1948.

The drive was taken by plate clutch and four-speed sliding-mesh gearbox to an underslung worm driven rear axle, the transmission line being inclined to assist in achieving a low floor level. Vacuum servo brakes were fitted and the frame, as well as being lowered between the axles in what has since become the orthodox manner, also tapered inwards at the front end.

The complete vehicle with the original standard open staircase body weighed about $5\frac{3}{4}$ tons unladen—appreciably less than modern lightweight double-deckers, and although the use of an oil engine would have added perhaps 3 cwt., it is interesting to note that the TD1 chassis weighed only $3\frac{1}{2}$ tons while the body accounted for the remaining $2\frac{1}{4}$ tons.

The body, despite its open staircase and the overall length of 25 ft., thus weighed slightly more than recent 27 ft. long lightweight double-deck bodies, so that the low weight, by modern standards, of the complete bus was not obtained by flimsy body construction. Although the interior finish was not up to modern standards in some aspects, interior lining panels were provided on both decks, being necessary to accommodate the full drop type of opening window fitted. The latter were of the spring loaded type and had a rather irritating tendency to gradually close when the bus was in motion.

The appearance was unlike that of any double-decker then in existence, although the "piano front" effect on the upper deck was widely copied during the next few years. It bore a remarkable resemblance, in this respect only, to some double-deckers placed in service by the Fifth Avenue Coach Co. of New York in 1925. The chief engineer of Leyland Motors during 1926-28, G. J. Rackham, had then just returned from holding a similar position with a firm associated with Fifth Avenue for about three years.

A six-wheeler of similar general design was also produced. This was the 'Titanic' TT1 of which only five (chassis numbers 75001-75005) were built. The relative importance of the four-wheelers may be judged by the fact that the 'Tiger' and 'Titan' chassis number series, starting at 60001 and 70001 respectively in 1927 had reached approximately 61937 and 72352 by the end of 1931, when the original series were superseded by new versions.

The TD1 had a wheelbase of 16 ft. 6 in. and the 'Tiger' TS1, TS2 and TS3 chassis which were the single-deck equivalent models were all mechanically similar to it and differed only in wheelbase and overall length. These variations were due to the differences in regulations regarding overall length, etc., which arose in different parts of the country prior to the 1930 Road Traffic Act. The TS1 was 27 ft. 6 in. long and had a 17 ft. 6 in. wheelbase: the TS2 had the same wheelbase but was shorter by 1 ft. 6 in. at the rear end, while the TS3 was of the same overall length as the TS2 but had a 16 ft. 6 in. wheelbase. All three models were available simultaneously except that the TS3 did not appear until 1930.

The TD1 as first introduced had a radiator which, while rather taller, was very like the PLSC in detail, with prominent external side gussets, and having a polished brass nameplate on the top tank bearing the legend "Leyland Titan" in script lettering. After the first few months this nameplate was replaced by the familiar aluminium type bearing "LEYLAND" in block capitals, and by 1929 the radiator shape was changed, the sides being slightly curved when viewed from the front and the side gussets having disappeared. The 'Tiger' radiator, however, remained of the multi-curved form throughout the 1928-31 period. For a period around 1929-30 the nameplates disappeared in favour of enamelled pictorial badges of which the current Leyland motifs are rather reminiscent.

In 1929 the PLSC and PLC models were replaced by new 'Lion' and 'Lioness' designs, designated LT1 and LTB1 respectively. Although having many detail parts in common with the 'Tiger' and 'Titan', these were new designs intended to inherit some of the simplicity of their predecessors. The LT1 had a new engine, of the same bore and stroke as that in the PLSC but of the same layout, apart from having four cylinders instead of six, as the T type. The remainder of the mechanical features were generally

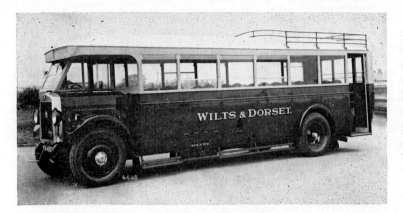

The *Lion LT1*, although having many design features in common with the Titan and Tiger, had few parts in common and consequently differed in appearance in many ways. This 1929 Wilts & Dorset vehicle—MW4597, chassis number 50326—is typical and its 31-seat Leyland body is of the style also fitted to Tigers in 1927-30 period.
[*Leyland Motors*

The original *Tiger*, although directly equivalent to the Titan TD1, had a type of radiator that clearly differed from both early or later TD1 radiators. This 1931 example of a *TS1*—SN5185 supplied to Clydebank Motors, which became Central SMT T71 in 1936—also had a 32-seat Leyland body of the standard 1930-32 style.
[*Leyland Motors*

similar to the 'Tiger' and 'Titan', but the transmission line was in the centre of the chassis instead of being offset to the nearside, and the frame was of the same width throughout its length. The radiator was accordingly of a more squat shape than that on the 1929 'Titan' which it otherwise resembled in detail. The wheelbase was 16 ft. $7\frac{1}{4}$ in. The general appearance, and that of the standard Leyland single-deck body as fitted to 'Tiger' and 'Lion' LT1 chassis during the 1928-30 period, was generally rather square-cut apart from the sloping windscreen. The new 'Lioness' model was fitted with the standard six-cylinder T-type engine and was sometimes referred to as the 'Lioness' "6", accordingly. The remainder of the LTB1 chassis, apart from the driving position was, similar to the LT1, and the radiator was of similarly squat proportions, although otherwise rather like the PLC in shape. It was, however, much further forward than on the PLC.

The LT1 and the much less numerous LTB1 chassis were numbered 50100 upwards, the numbers 50000 to 50093 having been issued to 'Leviathan' chassis. The LT1 model was in production for slightly over a year before it was superseded by the 'Lion' LT2, during which time about 700 had been built. The LT2 was basically a TS3 chassis with the same engine as had been fitted to the LT1, and having a radiator like the 'Titan' of the same period. This standardisation of parts caused only a slight increase in weight over the previous model. A considerably improved body for the 'Lion' and 'Tiger' models was introduced at the same time as the LT2 chassis, both interior finish and the more rounded exterior appearance being well above average for its time. The standardisation of 'Lion' and 'Tiger' ranges was carried a step further in 1931 with the introduction of the LT3, which was similar to the LT2 except for having a 17 ft. 6 in. rather than a 16 ft. 6 in. wheelbase, being thus a four-cylinder equivalent of the TS1.

Meanwhile, the standard Leyland coachwork fitted to the 'Titan' chassis had also undergone development. An enclosed staircase version appeared by early 1929, although open staircase examples were built to order until 1931. The former was one of the first double-deck designs to have a rounded rear dome, and the rear view had a much more sleek appearance than most of its contemporaries. Other alternatives included the seating capacity, which was invariably 48 or 51, with 24 or 27 respectively on the upper deck. The latter capacity was also standard for the first normal-height Leyland-built and overall length of 27 ft. 6 in. was standardised for the 'Lion' and 'Tiger' models. The 'Titan' wheelbase continued to be 16 ft. 6 in. with length now up to 26 ft.

With the introduction of the new range the chassis numbering system was changed, and all chassis built in the main factory at Leyland, regardless of type, were numbered in a single series commencing at 100. However, as is usually the case under such circumstances, production of the previous models did not cease immediately, and LT3 'Lions', for example, were built with chassis numbers at

This *Tiger* (left) is believed to have been the first TS4 and the first Leyland to have a chassis number in the series running up to 179xx that was used for larger models in the 'thirties. Thames Valley 239, RX9307, built in November 1931, had chassis number 100; the Brush 28-seat coach body was built to BET Federation design.
[Brush]

bodies for 'Titan' chassis, known as the Hybridge type and introduced in 1930. A number of other coachbuilders, including Massey, Northern Counties, Short Bros., Vickers and, possibly, Weymann also produced bodies of standard Leyland design during this period.

For 1932 a new range of 'Titan', 'Tiger' and 'Lion' models was introduced, designated TD2, TS4 and LT5 respectively. The principal differences from their predecessors were the provision of fully-floating in place of semi-floating rear axles, so that the driving shafts were relieved of the weight of the vehicle; three-servo, in place of single-servo, brakes and generally stronger construction to allow for the trend towards heavier bodywork. The bore of the 'Titan' and 'Tiger' engines was increased to $4\frac{1}{4}$ in., giving a capacity of 7·6 litres.

As the position regarding overall lengths of single-deckers had been clarified by the 1930 Road Traffic Act, a wheelbase of 17 ft. 6 in.

The early post-war view of Caledonian 246-WW7863, Chassis number 70343—a *Titan TD1* of the 1928-29 period, shows the type of replacement radiator produced from 1937 by the Coventry Radiator Co. for the 1928-32 Tigers but which was often to be found on Titan, Tiger, Lion LT and Lioness LTB models of that period.
[Alan Townsin]

least as high as 712, several hundred numbers after the first LT5 as well as the end of the old series around 51769.

The appearance of the new types was not greatly altered from that of the previous models. However a number of detail changes which provide useful aids to identification were made. The fully-floating rear axle had a considerably larger hub than the semi-floating type. Another guide is the front dumb iron design, a feature which is an important help in identifying most pre-war Leyland models, incidentally. On the 1928-31 models, the front springs were anchored at the rear ends, with the shackles at the front ends, whereas on the 1932 and all later models the more usual practice of anchoring the springs at the front was adopted. The accompanying sketches illustrate the resulting change in appearance, and also the variations in the shape of the dumb iron itself between the LT1/LTB1 type, which is more sharply rounded, and the remainder of the 1928-31 models. There was also a change in the dumb iron shape of the TD2, TS4 and LT5 models which occurred at about the end of 1932, when the "square" type was abandoned in favour of a "rounded" style rather similar to, but more massive than, the 1928-31 style, and, of course, without the shackles.

The radiators of the new models were also slightly changed, the 'Tiger' type having become a little deeper, so that the grille extended slightly below the starting handle shaft. However, the radiators are not a reliable guide to 1928-32 Leylands, since they can, in most cases, be interchanged and, quite often, in the course of time, this was done. In addition, from about 1937 a special replacement radiator, of similar outline, at the top, to the 1928-33 'Tiger', but considerably deeper, was made available from the

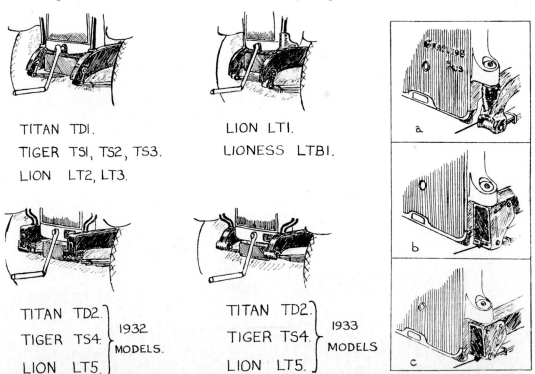

The variations in dumb-iron design on 1928-33 Leyland buses are shown in the four left-hand sketches; note that the 1928-31 models had the front spring shackles at the front, whereas the 1932-33 models had the springs shackled at the rear in the more conventional way. The three right-hand sketches show dumb-iron designs of 1933-42 Titans and Tigers—(a) TD3, TS6, TD4 and ST7; (b) TD5 and TS8; (c) TD7 and TS11, also TD6c with the addition of indicator board brackets; the Lion LT5A and LT7 had dumb-irons similar to, but not identical with, (a), and the same remarks apply to the LT8 and (b). The arrow indicates the place where the chassis number is usually found.

Coventry Radiator Co. It was originally intended for 'Tiger' chassis but has been fitted to all TD, TS, LT and LTB models of the 1928-33 period, although this necessitated modifications to the bonnet in the case of LT1 and LTB1 types.

The positive identification of 'Lion' and 'Tiger' models of this period can thus sometimes be a little difficult, especially as, apart from the LT1, the bonnet length of the former was exactly the same as the latter. However, on LT2, LT3 and LT5 types the clutch was sufficiently far forward to avoid the need for any protrusion of the cover over it into the saloon, whereas on the corresponding 'Tiger' models there had to be a "bulge," unless the floor level was unusually high.

The later TD2, TS4 and LT5 models, as well as having the modified dumb iron already referred to, also had a new type of gearbox, with constant mesh helical gears for third speed. This was, unlike some of its competitors, correctly described as a "silent third" gearbox, for it emitted a very subdued musical note when that ratio was engaged, the first and second speeds having a low pitched tone equally characteristic of these and later pre-war Leyland chassis. Both these changes did not affect interchangeability, however— the writer having seen several TS4 coaches with one "square" and one "rounded" dumb iron—and the chassis type designation was accordingly not changed.

Simultaneously with the introduction of the TD2, TS4 and LT5 models for 1932, Leyland returned to the manufacture of a smaller type of vehicle, a field into which the firm made a brief entry at an earlier date with the 'Leveret'. The new range formed the 'Cub' series, and they were built at the Kingston-on-Thames works which had been the home of the Trojan car with which Leyland were associated during the 'twenties. The type designations were, appropriately, based on the letter K, with suffix P for passenger and G for goods. The basic model was of normal control (bonneted) type, but forward control ("side type") models were also produced, in which case a prefix S was added to the designation, and left hand drive (bonneted only) models (prefix L) were built for export.

A six-cylinder petrol engine of $3\frac{3}{8}$ in. by 5 in. bore and stroke and 4·4 litres capacity was the standard power unit for the original 'Cub' range and this was notable for having side valves, being the only Leyland unit of this type designed after 1925. The remainder of the chassis design generally followed the practice of the larger Leyland models, with a "silent third" gearbox which justified that description, and fully floating rear axle. The brakes, however, were of Lockheed hydraulic type, servo assistance not being considered necessary on vehicles of the size in question. The appearance of the SKP models was very much that of a scaled down TS4 type 'Tiger' in general styling, and the KP series were similar, apart from the driving positoin, The 'Cub' model designation ran into quite a number of variants but the principal earlier passenger models were the KP2 of 14 ft. wheelbase, designed to seat up to 20, the KP3 (15 ft. 6 in., 24 seats), the KP4 (13 ft., 14 seats) and the SKP3 (15 ft. 6 in., 26 seats).

In 1934-35 a longer forward control model, of 16 ft. 3 in. wheelbase and designated SKP5, was built in small numbers. This was the 'Lion Cub', and it had a deeper version of the Cub radiator, so that the starting handle shaft projected through, rather than under, the grille. Also in 1934 a six-cylinder oil engine, of the same bore and stroke as the petrol unit was designed for the 'Cub' range. This was of the direct injection type and was to some extent, a scaled down version of the oil engine by then in production for larger Leyland models, having, of course, overhead valves. Models thus fitted had the letter "O" added to the designation, (KPO2 etc.).

Two new engines superseded the existing 'Cub' power units late in 1935. Although one was a petrol and the other an oil engine

Representative of early Leyland *Cub* models is this KP-series 20-seater with Brush body; numbered L392, LJ7519, in the Hants & Dorset fleet, it had chassis number 1659.

[Brush

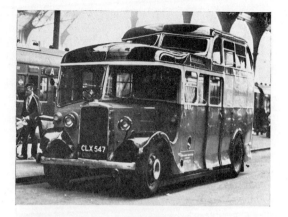

Later *Cub* models (*left*) had deeper radiators and heavily-slotted front wheels; vehicles supplied to London Transport also had non-standard wheel nut guards. C110, CLX547, in the LPTB fleet was a petrol-engined *SKPZ2* model with special Park Royal body designed for the Inter-Station service. Two views (*below*) of a *rear-engined Cub*—CR9, FXT115—supplied to London Transport before the war stopped production; note the hooded head lamps and white patches for wartime night driving.

[*London Transport*

they were designed to have as many features in common as possible. Both had push-rod-operated valves and were known as the "Light Six" engines. As compared with the previous units the bore was increased to $3\frac{1}{2}$ in., thus raising the capacity to 4·7 litres. and the oil engine was of the indirect rather than direct injection type. The Light Six-powered 'Cubs' (which, apart from the engines and general adoption of the longer radiator shell referred to above, were similar to their predecessors) incorporated the letter Z in their designations, O no longer being used to distinguish oil-engined models. From 1936 the KPZ1, KPZ2 and SKPZ2 replaced the KP2, KP3 and SKP3 respectively. In 1938 production of the SKPZ2 ceased in favour of the 'Cheetah' LZ3 (to be described later) and for 1939 vacuum-servo assistance was provided for the hydraulic brakes on the two bonneted models, which were redesignated KPZ3 (14 ft. wheelbase) and PKZ4 (15 ft. 6 in.) respectively.

Of non-standard 'Cub' models, the most outstanding were the chassis with rear-mounted oil engines supplied to London Transport. A prototype was delivered in 1938 and 48 out of 59 more ordered in 1939 were built before the war put a stop to further development.

Chassis numbers allocated to the 'Cub' models were in a separate series from those of the larger Leyland models. Both started at 100 in 1932, however, and thus there were two series of Leyland of about the same age with identical chassis numbers. The larger models were more numerous, and by the end of 1937 their series had reached nearly 18000 compared with almost exactly 9000 reached by the 'Cub' series in the same period. When the larger models began to be numbered from 300000 up at the beginning of 1938, the

THE LEYLAND STORY: 1926-1942

'Cubs' (and also the 'Cheetah' models, which had hitherto been in the "main" series) were numbered 200000 upwards, reaching over 202800 when the war stopped normal production.

Reverting to the story of larger Leyland models, the company had been working on the development of a six-cylinder oil engine suitable for the 'Titan' and 'Tiger' models since 1930. Several possible designs were tried before it was decided to employ the direct injection principle, using "flower pot" shaped cavities in the pistons and single hole injectors. The combustion chamber arrangement adopted gave a degree of smoothness of running well above average oil-engine standards, and indeed the claim was made, not without foundation, that it was virtually indistinguishable from that of the petrol unit apart from idling and low speed conditions. Fuel consumption was appreciably less than that of indirect injection engines then favoured by some other makers. The engine was notable at the time in being directly interchangeable with the equivalent petrol engines and in many respects the design was very similar, the valves, for example, being operated by an overhead camshaft. As originally put into production in the winter of 1932-33, the bore was $4\frac{3}{8}$ in., giving, with a stroke of $5\frac{1}{2}$ in., a capacity of 8·1 litres, but after the first year or so the bore figure was standardised at $4\frac{1}{2}$ in., making the capacity the better known figure of 8·6 litres.

A limited number of TD2 and TS4 chassis were built with oil engines. However, during 1933, these models were superseded by the TD3 and TS6, respectively, and either petrol or oil engines were available as standard for the latter, which, although generally similar to their immediate predecessors mechanically, had a completely revised front end layout. The engine, radiator, and driving position were all moved relative to the front axle, with the result that the length from the front of the vehicle to the front bulkhead was reduced from 4 ft. $11\frac{1}{2}$ in. to 4 ft. 5 in., the space thus saved being available for passenger accommodation. The steering column was less raked, but the most obvious external change was to the radiator, which was considerably deeper and had parallel sides. The previous practice of fitting different radiators to 'Titan' and 'Tiger' models was discontinued. The first production TS6 models were delivered in the early summer of 1933. Titans of the TD3 type began to appear a little later, but both TD2 and TS4 types remained in limited production for several months, so that although the early TS6 had chassis number around 2659, TD2 models were supplied with numbers as high as 2950.

A coach body was developed for the TS6 chassis by the company. Although this never want into large scale production, the Ribble and East Yorkshire concerns being the principal customers, the design was notable for its pleasing proportions, the large windows giving a spacious effect.

Another important mechanical development introduced in 1933 as an alternative to the standard clutch and "silent third" gearbox, was the Leyland torque converter. This was based on Lysholm-Smith patents, and

Ribble Motor Services relied on petrol-engined *Tigers* of the 1935-38 period for most coach services until the advent of the Royal Tiger fleet in 1951; a 1935 TS7— No. 1421, RN7587—with Ribble-designed 31-seat body built by English Electric is seen here in Accrington.

[F. G. Reynolds

The *Lion* LT5A was, despite its designation, very different from the LT5 and corresponded to the Tiger TS6; apart from having a four-cylinder engine, it could be distinguished from the TS6 by the distinctive radiator shape with its sloping sides and slender shell. The vehicle seen here leaving the Works was the last of 110 with oil engines and Leyland composite-framed bodywork supplied to Central SMT and the associated Lanarkshire Traction Co. in 1934.

[*Leyland Motors*

provided a fully automatic transmission, apart from a manually engaged direct drive for what would normally be regarded as "top gear" conditions. The latter was controlled by a lever (replacing the normal gear lever) which had four positions: direct, converter, neutral and reverse. No clutch pedal was fitted. The "converter" position was normally employed for all conditions from the vehicle being at rest (in which case the converter acted in a similar way to a fluid flywheel) to travelling at about 20 m.p.h., at which speed the direct drive was engaged. When starting from rest the engine speed would rapidly rise to about 1,200 r.p.m. and then remain almost constant as the vehicle accelerated with the converter acting as an infinitely variable ratio drive until the speed for engaging direct drive was reached. Similarly, when climbing a hill the engine speed remained steady as the speed dropped due to the increasing gradient, and these characteristics gave a false impression of sluggish and rather fussy performance which was probably heightened by the free wheel incorporated in the converter drive. However, there was an appreciable amount of actual slip, and the energy thus lost reappeared as heat, necessitating large coolers on the side of the chassis.

Vehicles with this transmission had a small suffix "c" added to their type designation, so that a TD3 with torque converter became TD3c. On the actual vehicle, the words GEARLESS BUS appeared on the radiator grille, and the normal Autovac in the nearside of the bulkhead was replaced by a taller narrower version together with a header tank for the converter. Gearless Leylands became quite popular during the middle 'thirties· particularly among municipalities, but, with certain important exceptions, orders specifying torque converters became less common as the need for maximum fuel economy increased. In quite a number of cases conversions of Gearless vehicles to orthodox transmission were made.

The revision of the front end layout of the 'Lion', to bring it into line with the 'Tiger' and 'Titan', did not go into production until the winter of 1933-34. It was, however more extensive, for full advantage of the compactness of the four-cylinder engiine was taken to reduce the bonnet length from 4 ft. 11½ in. on the LT5 to 3 ft. 11¼ in. on the new model, which, rather surprisingly in view of the extensive redesign, was designated LT5A. The wheelbase was, at the same time, increased by 1 in. to 17 ft. 7 in., and the steering column became almost vertical. The radiator was of an entirely new design with a shell which, in side elevation, was extremely slender and, when viewed from the front, sloped inwards towards the top.

A 5·7 litre four-cylinder oil engine, of similar design and cylinder dimensions to the 8·6 litre six-cylinder unit was offered as a standard alternative to the petrol unit for the LT5A. Some 110 oil-engined LT5A models were supplied to the Central SMT Co., and its close associate the Lanarkshire Traction Co., in 1934. These had 32-seat Leyland bodies which marked the end of an important stage in the company's coachbuilding activities, for they formed the last large scale order for Leyland bodies of composite construction.

Mention must also be made of the LT5B chassis supplied to W. Alexander & Sons in 1934. These largely reverted to the LT5 layout, being designed with the possibility of conversion from four-cylinder petrol to six-cylinder oil in mind. Some, in fact, entered service in the latter form.

By the end of 1934, oil engines powered the majority of new Leyland chassis of the larger types, although a substantial minority, particularly independent operators and the coach fleets of such concerns as Ribble and Southdown, continued to favour petrol-engined chassis until 1939. An increased bore size, $4\frac{9}{16}$ in., was available both for four- and six-cylinder petrol engines of the larger type, and by 1935 the four-cylinder unit of this size, of 5·9 litres capacity, was standard for the 'Lion' chassis. The six-cylinder version, which at 8·8 litres capacity was among the biggest petrol engines offered for p.s.v. use in Britain, was mainly intended for fire engines, although it was apparently installed in a few buses.

The three-servo vacuum braking system on the larger Leylands was superseded early in 1935 by a vacuum-hydraulic system, in which the servo unit was coupled to a hydraulic master cylinder. The model numbers were accordingly changed from TD3, TS6 and LT5A to TD4, TS7 and LT7 respectively. The introduction of the new models was spread over a period, the chassis numbers of the earliest examples being around 5600 while small numbers of TS6 and LT5A chassis were built up to at least a year later.

The identification of these two series in not easy unless it is possible to see the Lockheed brake header tank in the cab of the later models or the front brake servo units of the earlier types. It is appropriate to mention two "red herrings" which can mislead one when examining these models. At approximately the same time as the change of series, a modification adding a square hole in the dumb irons to give access to the radiator mounting was made. However, it was *not* related to the model change. It is of interest to note that the chassis type and number were stamped on, or adjacent to, the nearside dumb iron on all pre-war Leyland models, but, on the 1933 to 1937 models particularly, the dumb irons could be, and often have been, removed in the course of overhaul and refitted to chassis other than those from which they were taken. Dumb irons of TD3, TD4, TS6 and TS7 models were all mutually interchangeable and those of LT5A chassis could be changed with those of LT7 models but not with those of 'Titans' or 'Tigers'.

The 'Titanic' six-wheel double-decker became a normal production model for the first time in 1935. It will be recalled that it first appeared in TT1 form as a six-wheel version of the TD1. The TT2 was similarly based on the TD2, the most famous, and probably the only, examples being the three famous vehicles operated by the City concern in London (chassis numbers 2288-90, AGH 149-151). The 1935 versions were the TT3 and TT4 of 17 ft. $1\frac{1}{2}$ in. and 18 ft. $10\frac{1}{2}$ in. wheelbase respectively, based generally on the TD4 four-wheeler. In 1937, when the TD4 was replaced by the TD5 (to be described later) the TT3 and TT4 were superseded by the TT5 and TT6 respectively. The later 'Titanic' models were never produced in quantity but limited numbers were supplied to Bury and Doncaster Corporations.

Rather more numerous were the 'Tiger' six wheelers designed for single-deck bodies. These were very similar to the long wheelbase 'Titanic' models except that, generally, only one rear axle was driven, in which case the designations were TS6T or TS7T according to the 'Tiger' model to which they were related (the "T" suffix standing for "trailing axle"). There were also 18 TS7D models, with both rear axles driven, supplied to the City Coach Co. who had an equal number of the TS7T type, while the Central SMT Co. had 19 TS7T 'Tigers' with Leyland metal framed bodies.

The company turned to the manufacture of metal framed, rather than wooden framed (i.e. composite) bodies during 1934, it having been known for some time that development work in this direction was proceeding, there being a temporary drop in the numbers of Leyland bodies built. At first, attention was concentrated on the double-deck type, and interchangeability of parts, with jig built construction, together with comparatively light weight and low cost were the main aims of the design. The original design was built in comparatively large numbers during 1935 and the earlier part of 1936, and although it was not entirely satisfactory at first, subsequent changes were so effective that in quite a number of cases a life of over 15 years has been obtained. Some bodies of this type remained in service at Portsmouth for 28 years. The pronounced vee-fronted effect on the upper deck and the waistrail mouldings were in keeping with the generally rather angular appearance, and the radiused window corners were noteworthy.

During the summer of 1936 a revised design of Leyland metal framed double-deck body was introduced. Apart from incorporating structural improvements following experience with the earlier examples of the previous type, the appearance was entirely new. The number of window bays between bulkheads was reduced from six, common to all previous Leyland bodies for 'Titan' chassis, to five, and radiused window corners were, at first, abandoned. The front end was given the unbroken curved profile then just beginning to come into favour. Indeed, so far as the writer is aware, this design was the first double-decker on which the cab front panel was made practically flush with the front of the radiator, and curved in plan view, so that, with the flush fitting windscreen, a particularly sleek appearance was obtained. This was something of a landmark in double-deck appearance. At the rear, a similar change was effected, although the lines were still comparatively upright. An interesting detail was the shape of the single window in the upper-deck rear emergency exit. This was of less than the full width of the opening and had sloping sides: it seems likely that its design, and possibly that of other features of the body, was related to the trolleybus bodies then being produced by the firm for London Transport.

Leyland trolleybuses were produced in quantity from about 1933, although earlier vehicles, basically conversions of petrol-engined chassis had been built. Space does not permit of details being quoted here, but four- and six-wheelers were built, the designations being in the TB and TTB series respectively.

Single-deck Leyland metal bodies were built for 'Lion' and 'Tiger' chassis between 1935 and 1939, although the quantities never rivalled those of the double-deck types. Details of window and frontal design varied over the period of production and a choice of front or rear entrances was available.

At the November 1935 Show the 'Cheetah' model was introduced. This was of similar layout to the 'Lion' but incorporated a large proportion of parts taken from the Cub' range, being designed as a lightweight vehicle. The engine was the 4·7 litre Light Six (petrol or oil) and the clutch, gearbox, rear axle and brake system were also based on 'Cub' units, although the change speed was designed to suit the full forward control layout. The frame and front axle beam were new, although the stub axle assemblies were of 'Cub' type. The appearance, at first glance, closely resembled the 'Lion', for the radiator bonnet, steering column and general front end layout were of 'Lion' type, but the 'Cub' type wheels and hubs provide an easy means of identification. The original models were the LZ1 of 16 ft. 9 in. and the LZ2 of 17 ft. 7 in. wheelbase. In 1937 the LZ2A model, with vacuum hydraulic brakes, superseded the LZ2. Shortly afterwards the LZ1 and the forward control 'Cub' type SKPZ2 were replaced by the LZ3, with a 15 ft. 6 in. wheelbase, also with vacuum hydraulic brakes. Further braking improvements were made on the LZ4 (15 ft. 6 in.) and LZ5 (17 ft. 7 in.) models, introduced late in 1938. The Ribble concern had a fleet of over 300 petrol-engined 'Cheetahs', and large batches were placed in service by some of the SMT group companies, including some with oil engines and bodies having the radiator hidden by a full front. The long wheelbase petrol 'Cheetah' was an attractive commercial proposition when fuel costs were less important than they are now, and it had a smooth and lively performance, although one or two operators were inclined to take unfair advantage of the body space available by fitting heavier bodywork than was intended.

During 1937 a fleet of 100 'Titans' with Leyland bodies was supplied to London Transport. Both chassis and bodywork of these incorporated many special requirements, although they were based on the standard TD4 model. The chassis modifications included worm and nut (in place of Marles) steering gear and a revised frame to which reference will be made later. The body conformed to LPTB standards in equipment, although the shell was of standard design. Until their withdrawal from service around 1954, all the 100 remained in service: a remarkable tribute to their original design

The *Cheetah* had the same radiator and general proportions as the Lion LT5A, LT7 and LT8 models, but its lightweight construction was belied by the Cub-type wheels. Ribble had over 300 petrol-engined Cheetahs, including many with spacious 30- or 32-seat bus bodies like 2103, RN 8668, seen here at Whalley; this is a Burlingham-bodied example and was converted to diesel operation in 1948.
[F. G. Reynolds]

and manufacture as well as the maintenance they received, although similar results with pre-war Leyland metal bodied 'Titans' are by no means uncommon. It is interesting to compare the body weight, approximately 2 tons 10 cwt., with that of more recent lightweight double-deckers, for although the external dimensions have increased, the pre-war Leyland body had none of the "tinny" effects of some present-day products of other makes.

At the November 1937 Show, the frame modifications referred to above, together with other changes, became standard and the 'Titan', 'Tiger' and 'Lion' models became types TD5, TS8 and LT8 respectively. The front dumb iron design forms the simplest means of identifying these models, and this accounts for some confusion over the special 'Titans' supplied to the LPTB in 1937 which, although of TD4 type, conform to the TD5 design in this respect. The change was from the more rounded type of dumb iron fitted to 1933-37 models to the "square" type of the TD5/TS8/LT8 models. The first TD5 and TS8 models had chassis numbers around 13600 but the last LT7 chassis had numbers as high as about 17400.

An alternative version of the 'Lion' model, designated LT9, was also put into limited production at the same time. This was basically a TS8 chassis with four-cylinder engine, the appearance and bonnet length being that of the 'Tiger'. This enabled a more spacious cab to be provided where maximum seating capacity was not required, and, as with the LT2, LT3 and LT5 types, the fact that the clutch was entirely under the bonnet removed the need for a large cowl inside the body, even with a low floor line. In addition there was a change to the wheelbase of the standard 'Lion', which became 17 ft. 8½ in. on the LT8.

A more revolutionary design introduced at the same show was the twin steering 'Gnu' six-wheeler. The original model was the TEP1, of which only three examples were built, although five more, which were basically similar to the Steer goods chassis, being of type TEC2, were built in 1939. A further model on similar lines, in some respects, was the 'Panda', which had an underfloor-engine in a twin-steering chassis, of which one example was delivered in 1940.

The 'Panda' was not the first underfloor-engined model, however, for the company built the London Transport TF class vehicles. A total of 88 of these vehicles of which the Leyland designation was 'Tiger' FEC, were built; one in 1937 and the remainder in 1939.

On the other hand the large normal-control coach had not died out, for a bonneted version of the 'Tiger', known as the 'Tigress' was built for Southdown, six being supplied in 1936, and a fleet of similar vehicles was exported to Estonia.

It is interesting to recall that the company introduced new six-cylinder petrol engines for the 'Tiger' and 'Titan' models as late as 1938, reflecting the substantial minority of operators then still favouring this type. Certainly, from the passenger's viewpoint, the petrol 'Tiger' offered extreme refinement especially when the engine was idling, for it could be quite inaudible inside the vehicle. The new engines, known as the Mark III

The now familiar Bedford VAL layout was anticipated by the Leyland *Gnu*; the City Coach Co's G1 FGC593 chassis number 16769—was one of the three *TEP1* models built, and entered service in 1938 with the modern-looking 40-seat Duple body.

[City Coach Co.

Leyland's first under floor-engined, model was the *Tiger EFC*, of which 88 were built for London Transport before the war; the great majority of the LT vehicles was used for Green Line services, including TF16—FJJ617 —seen here equipped for use on the pre-war route X to Romford, via Eastern Avenue.

[London Transport

Representative of the standard Leyland metal-framed doubel-deck body for the Titan chassis in its 1936-38 form was Accrington Corporation 89 — CTC217 — a 56-seater supplied in 1937; the basic styling of the upper deck front remained unchanged until Leyland body building ceased in 1953.

[Leyland Motors

type, differed from their predecessors in having push-rod-operated overhead valves, and were of 4 in. or $4\frac{5}{16}$ in. bore and $5\frac{1}{2}$ in. stroke, giving alternative capacities of 6·8 and 7·9 litres. The 6·8 litre type was nominally standard, although the Bournemouth TD5 type 'Titans' delivered in 1939 had the 7·9 litre unit, giving them a very lively performance. Many operators continued to put overhead camshaft petrol-engined vehicles into service.

The standard metal body for the 'Titan' was improved by the introduction of a more rounded rear dome and radiused window corners in 1938. During 1938-39 an interesting variation of the standard body was built in some quantity for Manchester Corporation. These vehicles incorporated revised window outlines and other modifications to suit the operator's specification and conform to other bodies in the same fleet. They were, however, based on standard Leyland frames as far as possible.

It had evidently been intended to introduce two new models, the 'Titan' TD6 and 'Tiger' TS10, during 1939; mention of them is to be found in standard reference books. However the first vehicles incorporating the principal new features were the 85 torque converter 'Titan' chassis for Birmingham City Transport; chassis numbers 300528-612, supplied late in 1938. The order for a further batch of 50 (303155-204) specified Leyland bodies. This design incorporated many Birmingham features, including a straight staircase (as on an earlier batch of five Leyland bodies on TD4c chassis for the same operator). double skin roof and other fittings, but as with the Manchester vehicles, modifications to the standard body frame were kept to the minimum. The resulting bodies were, in the opinion of many, among the most handsome ever placed in service by Birmingham. The chassis also included many of the operator's special requirements, and accordingly it was decided that, to avoid confusion, the designation TD6c would be confined to the Birmingham chassis, and that standard models incorporating the new features would be called TD7 and TS11. Thus no TS10 chassis were built.

The principal new features of the TD6c, TD7 and TS11 models were fully flexible engine mountings (previous models having had a semi-flexible arrangement providing insulation rather than freedom of movement), worm and nut steering, a slightly raised driving position, triple-servo brakes, longer road springs, and on the 'Titan' 16 ft. 3 in. wheelbase in place of 16 ft. 6 in. and rear platform extensions formed as part of the main sidemembers rather than being bolted on, as hitherto.

Identification of these models is possible by the type of dumb iron, the depth of the front portion, and especially the lower strip (on which, in both cases, the chassis number is stamped) being less than on the TD5, etc. Other points included a smaller front hub cap, resembling the post-war type, and the enlarged top bearing required for the worm and nut steering. The bonnet was also devoid of holes or louvres, although this has sometimes been altered subsequently.

The standard Leyland body for the TD7 was almost identical with that fitted on TD5 models, except that the wider double emergency windows, as fitted to the bodies supplied to Birmingham on TD6c chassis, had become standard.

The TD7 model came into production in the latter part of 1939, but the TS11 did not appear until 1941. The earliest TD7 chassis numbers in the writer's records—actually TD7c models—are 303230-254 (Bolton) but these were the balance of an order for TD5c chassis, and the 304400 region is nearer the mark as a general guide. The last TS8 models had numbers over 306600, and among the later examples to be built were included about 100 for the SMT group of companies with modified driving position to allow the fitting of 39-seat Alexander bodies. The steering column was in a similar position to that on the LT8 model and the bulkhead was moved forward about 6 in., with the rear of the cylinder block, which protruded through it, covered by a cowl.

Leyland bus chassis production ceased for the duration of the war during the winter of 1941-42, the last to be turned out being about 220 "unfrozen" TD7 and TS11 chassis which were allocated to a wide variety of operations by the Ministry of War Transport.

The Second Deck

The second deck on buses was an English invention to meet the traffic problems of the 1851 exhibition in London. Although condemned by Continental busmen as being clumsy and time-wasting double deckers are now popular in Spain and Germany; there is a large fleet in Vienna and several Italian undertakings have experiments in hand. France is the latest adherent to the idea.

In the summer of 1966 the Paris Transport Board produced this 53 seat 110 passenger double-decker, intended as the prototype of a large scale experiment.

THE SECOND DECK 95

Based on a Federal lorry chassis, this Aerocar is a specially-built job in the fleet of Autocares Gelabert of Barcelona. Passengers have a completely unobstructed view because the driver is located in a racing driver's stance above their heads. Controls pass down one of the slender windscreen pillars.
[C. F. Klapper

A rural half double-deck bus on Pegaso chassis seen at Vitoria in Spain; the rear part of the roof is for luggage.
[C. F. Klapper

An earlier Paris double-decker at the end of the 'fifties was this Berliet operated in conjunction with the Bateaux-Mouches pleasure ships on the Seine.

[C. F. Klapper

In Germany the double-decker has a firm hold in Berlin and a few other places such as Travemunde, but elsewhere the 1½-deck high-capacity bus is considerably used. In the picture is one operated by the Cologne transport undertaking.

[C. F. Klapper

Alexander King

Alexander King